The Rage to Survive

THE RAGE TO SURVIVE

by Jacques Vignes

TRANSLATED FROM THE FRENCH BY
Mihailo Voukitchevitch

Illustrated with photographs, maps, charts, and original drawings by Ardea

Preface by Alain Bombard

WILLIAM MORROW AND COMPANY, INC.
New York   1976

English translation copyright © 1975 by William Morrow and Company, Inc. and Granada Publishing Limited

Originally published in French under the title *La rage de survivre* by B. Arthaud, Paris, copyright © 1973 by B. Arthaud.

All rights reserved. No part of this book may be reproduced or utilized in any form or by any means, electronic or mechanical, including photocopying, recording or by any information storage and retrieval system, without permission in writing from the Publisher. Inquiries should be addressed to William Morrow and Company, Inc., 105 Madison Ave., New York, N.Y. 10016.

Printed in the United States of America.

1 2 3 4 5 79 78 77 76 75

Library of Congress Cataloging in Publication Data

Vignes, Jacques (date)
  The rage to survive.

   Translation of La rage de survivre.
   1. Njord (Sailboat) 2. Shipwrecks—Mediterranean Sea.
3. Survival (after airplane accidents, shipwrecks, etc.) I. Title.
G530.N68V5313   910'.45   75-23312
ISBN 0-688-02992-2

BOOK DESIGN: HELEN ROBERTS

To Catherine and Lucien, whose story this is

Preface

"The Book of Courage and of Fear!" What a pity that title has already been used by Rémy for the second volume of his memoirs. It would make an admirable subtitle for the book you are about to read.

What an extraordinary adventure of survival at sea. What clarity in the recounting of events. I was spellbound by this narrative and would gladly repeat to Catherine and Lucien the comments of a reader: "Reading your story, I was so frightened that I was afraid you would perish before the end!"

I sought the honor of writing the preface to this exemplary book. It is a priceless testimony. How many shipwreck victims indeed have had the luck and the talent to be able to record hour by hour their dramatic odyssey.

Personally, I found in these pages grounds for admiration, satisfaction, hope, and concern.

Admiration for the feat: to have survived against the sea, the wind, fear, thirst, hunger, cramps, misery, and despair. Let no one scoff at this enumeration. All those monsters *had* to be fought and overcome. Life was at stake.

Admiration also for the sharpness with which every blow of fate, every disappointment, every hope of rescue was felt, analyzed, and described.

Admiration for the utter sincerity of the narrators, for the unembellished truthfulness of their every word, without fear of appearing to despair.

Satisfaction because I find two cardinal principles of my own personal experience confirmed by this narrative. It was *spirit* that saved Catherine and Lucien. Every time their spirit faltered, death drew near. Every time it rose again, that fearful specter retreated.

You will have the great joy of seeing hope live, vacillate, spring again, and triumph in the narrative of our heroes.

The second point (Lucien will forgive me if I underline it) is the preeminence of spirit in woman as opposed to man. I have often remarked that women, those exceptional beings, quickly dispirited in the face of imagined danger, are superbly strong and courageous confronted by actual peril.

Lucien, are you sure you would have survived without Catherine? I have met Catherine, I have not yet (May, 1973) met Lucien. Catherine was not boastful, yet despite her calmness, her modesty, her freshness of soul, in the narrative I expected to find her a pillar of strength against which adversity was powerless.

What an admirable team, this man and this woman! How clearly this young woman demonstrated the necessity of couples, and women's lib be damned!

My concern arose over the failure of the inflatable raft. The only explanation I can find is its small size. A larger minimum surface than the one you had, Catherine and Lucien, is needed to withstand breaking seas. Even so, the crying out of your harsh experience will serve to remove from circulation ineffectual and dangerous rafts.

My hope is that you, young woman and young man, have shown that the will to live can make death withdraw its face.

Forgive me if I thank heaven for having sent you so dangerous, so ineffectual a device, for it allowed you to surpass yourselves. You owe your life neither to men nor to the elements but rather to the noblest part of your being, your capacity for thought.

While reading this book I felt that you, Catherine and Lucien, were my children.

ALAIN BOMBARD [*]

---

[*] Author of *The Bombard Story* (Penguin Books, 1956), recounting his single-handed voyage across the Atlantic in a rubber dinghy. During his sixty-five days afloat he lived off the sea—eating fish and plankton, drinking rainwater and seawater—and arrived in Barbados in very good health.

Contents

Preface   9

I   THE STORM
   1. *Njord*   17
   2. First Conversation: Heading for Disaster   37
   3. As Far as One Can Go   48
   4. Second Conversation: To the End of the Nightmare   68

II   SURVIVAL
   5. Turnabout from the East   83
   6. Third Conversation: The Wait   98
   7. When the End Has Come   111
   8. Fourth Conversation: Hunger   123

III   THE OUTCOME
   9. Land   135

10. Fifth Conversation: Fog	153
11. The Last Postponement	164
12. The *Abel Tasman*	179

Commentary   190

I. THE STORM

1. Njord

When the staysail ripped, early in the afternoon, *Njord* had been running downwind since 2:00 in the morning. A howling northwest wind was trailing long whitish streaks across a crossed gray-green sea. Despite their 10- to 12-foot troughs, the waves had not yet become really worrisome. The sky was leaden. It was an odd situation. In the gulf of Lyons, the tramontane or cers winds usually blow in clear weather, with the horizon etched beneath the preternatural calm of a white sun. That morning, when a bit of misty rain, like a Breton drizzle, briefly mingled with the spray, Lucien had decided the disturbance was a local one. The squall ought to blow over soon.

Njord was a 26-foot gaff-rigged steel cutter, a Colin Archer double-ender, weighing, empty, about seven tons. She was a heavy boat, further freighted with gear shipped for a long cruise—the Canary Islands, the West Indies, and after that, wherever wind and wish might bid. Sailing her were Lucien, twenty-five, and Catherine, who was nineteen.

Hailing from Luxembourg, Lucien had dreamed of blue-water sailing since he first started racing skiffs in his teens. The dream began to come true on various cruises. He delivered boats from the Atlantic to the Mediterranean, and crewed on big yachts belonging to those wealthy summer sailors who idly ply between the Riviera, Corsica, and the Balearic Isles.

It had been two years ago, in the Balearics, that Lucien had bought *Njord,* her old hull still in good shape under a shroud of rust. Quickly patching up the rigging and the sails that were left, he made it to Beaulieu, on the French coast. There, moonlighting between two bread-and-butter jobs, he had turned to the task of refitting his boat.

Catherine was the daughter of a career international civil servant stationed in Geneva. In 1971, she came to Nice for a year of higher mathematics, and she too made Beaulieu her home port. At the end of three months, she decided to quit school. She simply had to lead her own life—but still had to find something to do.

At nine years old, while living at Addis Ababa, she had already become an accomplished rider and dreamed of ranging the world on horseback. Later, sailing with her family on a freighter taking them from southern Africa to New York, a storm at sea gave her a new goal: to become a sailor. She learned to swim in Paris and ski in Switzerland, meanwhile leading a fairly happy life as a studious teen-ager. But now, in January 1972, the shaky balance had broken down. She had to find a new outlet. Roaming the quays of Beaulieu as she mulled over the matter, her eye lit on a rusted hull, just waiting for someone to scrape her.

A friendly redhead with a beard was also gazing at

the boat—his boat. He was planning to go off to sail the world soon. The redhead's name was Lucien. Catherine saw him a few more times, then found herself scraping rust at his side. Together, the job progressed well. By spring, their relationship had flowered, *Njord* was back in the water, and almost ready to sail.

Before leaving, however, they had to earn enough money to finish the refitting. A gleaming cabin cruiser solved the problem. Lucien signed on for a stint as skipper, and Catherine as cook. By the end of August, however, a disagreement with the owner arose, and they again found themselves on the beach. Why not move up the big day by a few months, then? To be sure, there were still a few things that had to be done on board. In particular, *Njord* had no self-steering rig; but wouldn't it be better to do the job somewhere else? The idea of spending another winter in a port they knew only too well, going through the motions of the same daily routine, was just too much. In her present state, *Njord* seemed ready for her first passage.

How far would they go? To Gibraltar, perhaps even to the Canary Islands. Taking just enough time to collect the essentials, they set sail on September 11th in a light northeast breeze, reaching Porquerolles on the 12th. Spending the night there, they left at dawn on the 13th for the Balearic Isles, a possible port of call.

The weather report was good: northeast winds 20 to 25 knots, perfect for the passage. Lucien knew the gulf of Lyon, but he did not like it, since he always seemed to run into some tramontane or other there. Moreover, the first day's sailing was unexpectedly calm—too calm, in fact. The northeast wind put in an appearance, but was so light that *Njord* ghosted along on a flat sea, barely covering twenty miles in a day's sail. The wind fell with evening, as a northeast swell began to run.

Still very close to the coast, they could see the lights of Toulon. Would it not be wiser to turn back, and get a quiet night's sleep somewhere, rather than rolling helplessly, and risk a collision with one of the steady stream of passing freighters?

Opting for discretion, they set about mounting their outboard motor on its bracket.

Unfortunately, just as they were about to get under way, *Njord*'s stern dipped into a swell, swamping the motor and making it temporarily useless. With the boat rolling in the darkness, dismantling and drying the carburetor was out of the question.

They could only wait, and hope the wind would rise during the night. At about 9:00 P.M., a light northwest breeze came up. Lucien and Catherine were undecided. Should they take advantage of the wind to reach land, or make for their goal? It seemed a bit silly to back down as soon as things got tough, in the face of their first threat of heavy weather, when they had decided to take on the challenge of the open seas.

The high swell was a sure sign of a mistral or tramontane blow. That the forecast still predicted northeast winds 20 to 25 knots next day made no difference. But after all, mistrals or tramontanes are not the end of the world. Better too much wind than more of those last dull days. Accordingly, they set off on a reach to the southwest, steering 230.

Toward midnight, the wind dropped, leaving *Njord* bobbing like a cork on the deepening swell. Thrown from side to side, her rigging screeched all night long, while gear rattled below. They could hardly move around without danger of losing their balance. At dawn, the northwester came up again, this time quickly reaching force four to five.

During the day, the wind helped them eat up the miles. They took turns at the helm, three hours at a stretch. At sunset the wind showed no sign of slackening. By midnight, in fact, Catherine found she had to hold the tiller with both hands. Even then, she could not always control it. She would be helplessly dragged to leeward as the boat luffed up, sails shivering in the wind.

Taking a reef in the mainsail eased the helm a little. Forward, *Njord* was carrying only a well-worn cotton staysail; the jib had been handed, so as not to spend the night with too much canvas.

Toward 2:00 A.M., the situation worsened still further, and they decided to run under bare poles until daylight. The very crossed sea threw them around, and Lucien rigged the staysail again at dawn, hoping to ease the boat.

As he was about to do so, he briefly considered trading the old sail, which was still hanked on, for a spare that was stronger, or even a storm jib. He had not slept much the last two days, though, and poorly at that, because of the boat's brutal motion. Fatigue was beginning to take its toll. Raising the old staysail was the easy way out, and he took it. Seven hours later, the sail burst along one of its panels and promptly ripped to shreds.

They were getting more and more tired. Steering in the choppy sea was extremely tedious. Moreover, there was no way to fix hot drinks, like coffee or tea. The boat was rolling so badly, the gas stove could not be lit. They had hardly eaten since the night before, and Lucien especially was suffering from that insdious form of seasickness that does not really make you nauseous, but keeps you from eating by making food unthinkable. One thing was becoming obvious. They must rest, heave to one way or another, let the boat take care of herself, and stretch out on their bunks, warm and dry. They were soaked despite their oilskins, and the wind had become colder as it freshened.

To be warm and dry, perhaps to sleep, even. The thought seemed especially appealing because they felt no real anxiety. Lucien trusted his boat implicitly. After all, had he not faced nearly identical wind and sea conditions when sailing her from the Balearics to the Continent two years before?

In those days, breakdowns had been his chronic worry. Since then, everything had been repaired and strengthened; nothing was likely to fail. But he was furious, raging against the damned gulf of Lyons, where the elements seemed to cut loose every time he poked his bowsprit in. Why hadn't he followed up on his initial idea, the day before yesterday evening? Use those three hours of providential wind to put into Toulon or Embiez, and be as snug as you please.

Catherine, for her part, was not worried either. In fact, she was rather excited by the spectacle of the boiling sea and the wind howling in the rigging—finding a kind of calm amid the vortex of forces run amok. This was her first storm aboard such a small boat, and it brought back memories of that other storm, the one she rode out aboard the freighter bringing her to the United States. Though she had been expressly told not to, she had slipped outside and made her way on deck, the better to see the waves. For a while, her family thought she had been swept overboard by a wave. Then they found her clinging to the rail, soaked, freezing, and in ecstacy.

To steady *Njord*, Lucien set a sea anchor from the stern, on about sixty yards of line. Why the stern? "I was convinced," he explained later, "that a double-ender's stern would meet seas as well as her bow."

This done, they went below, undressed, and stretched out on the bunk. Sooner or later, the storm would surely blow itself out. It was about three in the afternoon.

Steadied by her sea anchor, *Njord* was riding well. Now that she was dead in the water, however, or at most drifting very slowly, she offered more resistance to the waves, and her motion became more brutal, particularly when seas struck her amidships. One of these blows sent Catherine crashing through the lee board on her berth.

At the same time, the outside noise, amplified by the metal hull, rose to a deafening roar. The boat seemed to be groaning, as if each of its ribs was being wrenched by the continued battering. A demonic concert was tuning up, a cacophony of high shrieks and low rumbling, punctuated by the rattle of the rigging, the howling gusts, and an occasional tremor that would shake everything.

Assailed by the din, Lucien and Catherine could not really relax. Their stomachs would knot at each new noise, or whenever the row got particularly deafening. At times, everything lashed on deck would start vibrating uncontrollably. There was an inflatable life raft along with a dinghy, a boat hook, a boom, and the outboard motor that had declined to run some forty-eight hours before. The entire lot seemed to be dancing a wild saraband, warming up for a final leap into space.

Lucien and Catherine were discussing taking a turn

on deck to check the lashings when their whole world capsized. *Njord* went down on her beam ends, probably below the horizontal, since Lucien went flying to the cabin top, breaking one of its wooden panels. He fell back on top of Catherine, but neither was hurt. At the same time, all the stores and gear stowed to port came crashing loose with a clatter of breaking glass. A few seconds later, when the boat abruptly righted herself, everything on the starboard side was scattered about as well. It was 5:30 P.M.

The pair rushed on deck, trampling tea and marmalade underfoot. *Njord*'s main structures seemed intact, but chaos reigned below, as if someone had spitefully strewn everything helter-skelter. One distressing detail: both flashlights were shattered, and the matches soaked; there was no way to light the gas stove or the hurricane lamps. They had planned to bring lighters, but forgot them in the haste of departure.

Outside, the fury of sea and wind had redoubled. The sky was so leaden, it looked like dusk. It was cold. Unusually low temperatures had been forecast for the 15th. This time, the prediction had come true.

The cold hit them as soon as they left the cabin, since they were hardly dressed. Lucien had hurriedly put on a life jacket, but was otherwise practically naked. Catherine was wearing only a bathing suit and a blouse. Moreover, foam and spray soon soaked them anew.

Still, going below to get dressed was out of the question. They had to act fast: at any moment, another crossed sea could viciously knock *Njord* down again, perhaps even roll her completely over. Even if it did not sweep them overboard, this could be very serious. They had to free the boat by bringing the sea anchor in, and get her underway again; then, taking turns at the helm, seize the first chance they got to go forward and raise a storm jib.

The day was drawing to a close. The wind might slacken at sunset. If they could hold on for just a couple of hours more, perhaps things would get better. They could heave to again, long enough to rest, before resetting their course for the Balearics. While awaiting the lull, however, they would have to struggle, meeting every wave, foreseeing every danger, keeping the smallest detail constantly in mind. With Catherine at the helm, Lucien hauled the sea anchor in without any trouble, then spelled her. Chilled through, Catherine went below for some clothes.

In the cabin, with its tiny portholes, she could hardly see. The lockers had spewed all their contents in the capsize, and Catherine had to grope for dry clothes among the odds and ends. She came across a blouse, a sweater, and a pair of cotton pants. She wanted to dress more warmly, but the thought of Lucien outside, waiting for his chance to put on some clothes, made her hurry.

Besides, she felt uneasy in the dark, narrow space, where motion was as brutal as it was unexpected. She hardly knew where to step, so many odds and ends littered the cabin floor. The feeling grew of being caught in a trap, making her even more uneasy.

Managing to put on clothes was an exercise in acrobatics. No sooner would she let a handhold go, than she might be suddenly thrown any which way, and perhaps hurt. And above it all, there was the unceasing din, the noise that would swell from whistling wind and rumbling seas in the cockpit to pandemonium below, roaring as it had a short while before, when *Njord* was knocked down.

Despite her best efforts, it took Catherine a half hour, resting from time to time, to get ready. After putting on her pants, blouse, and sweater, she managed to find a set of oilskins and a life jacket. It was not the best, but would have to do for the night . . . unless the wind dropped, the sea calmed a little, and they could quickly clear the bunks, slip under the covers, and sleep to their hearts' content.

Once outside in the open air again, free of the oppressive closeness of the cabin, Catherine went to relieve Lucien at the helm.

The boat was riding well. While she tended to yaw when hit aft by a wave, it was not hard to keep her in the eye of the wind. The repeated drenchings that hit the helmsman were more annoying. The cold quickly penetrated the oilskins, as water seeped down the collar and up the sleeves, soaking into the woolens.

Meanwhile, in the cabin, Lucien was having the same difficulties Catherine had faced a few moments before. Unable to stand upright, he groped for his clothes in the gathering darkness. Crawling along the floor, he clutched at anything he found that resembled what he was looking for. He had been outside for nearly an hour without any clothes. Now, chilled to the bone, he was in a hurry to get warm again. Despite his haste, he wound up doing better than Catherine, emerging with a shirt, two thick sweaters, wool pants, oilskins, and a life jacket.

At about 7:00 P.M., Lucien took the tiller again, and settled himself to port. Catherine went to the forward end of the cockpit, to sit against the starboard cabin bulkhead.

She was hoping to get a bit of sleep, or at least to doze while waiting to spell Lucien. She preferred to stay outside, at the risk of getting wet and cold. The memory of the fear that had gripped her in the cabin was still fresh in her mind. For the moment, she would just as soon not go below.

While trying to make herself comfortable, she watched Lucien carefully handling the tiller, apparently controlling the boat's motion without undue effort. Aft, a tireless procession of waves pursued them. Some were rounded, almost friendly. Others broke endlessly. The frothing masses could be seen coming from afar, and usually missed *Njord* by a few cable lengths. Their white flanks could be followed a long way, catching the dying light.

Others, however, had to be reckoned with: threatening cones capped by shining, flowing crests, bizarre vertical masses of water, ever on the point of imminent collapse. Those waves, most often moving erratically across the general direction of the basic swell, were the most frightening. They looked as if they were born of the collision between two waves belonging to separate systems. When they had met, each rose to submerge its rival. When neither gained the upper hand, they abruptly joined forces, merging their power into a single peak moving high above the parade of lesser parallel crests, brutally shattering their easy regularity.

The abnormal mutations were short-lived. Driven by the wind, the wandering colossi soon collapsed, or were absorbed. In either case, they were destined to disappear. No sooner had one vanished, though, but another rose in its stead, roaring along in the same headlong rush. At times, two, three, or four erupted together, giving the sea its most awesome visage—that of an irrational, incoherent power, whose next blow would fall where it was least expected.

From her front row seat, Catherine was fascinated by the stupendous ballet all around, seemingly danced for her alone, heedless of *Njord,* sailing along at a fairly even rhythm. Driven by a wave, the boat would surge powerfully forward, seem to pause for a moment, balanced at its peak, then slide heavily into the trough, awaiting the next thrust.

She would not be able to watch this incredible spectacle much longer, however. Already, around the glistening sheets of foam, the sea was beginning to take on the slate-gray color of dusk that seems to lengthen those stillborn days when the sun stays hidden and the roaring wind blasts out a space between boiling sea and leaden clouds, keeping them from crushing that fragile space between, in which movement was still possible. In a few moments, night would fall, and in that dark night, no spectator would see the unfolding of the ballet. Unless, by chance, the demon dance came to an end, and *Njord,* relieved, took her course again, across a tranquil world.

Catherine was beginning to feel that the seas, after all, were like those blusterers whose threats were more frightening than dangerous. She was about to relax her vigil of the shifting, watery jumble, when a scream froze in her throat. There, barely thirty yards behind *Njord,* a monstrous wave was approaching, a veritable wall of sea-green water, its crest dangerously curling. She wanted to shout to Lucien, but couldn't. The wave was unlike any they had seen before. What was going to happen? It was unbelievable.

She just had time to grab the main sheet and take a turn around her hand before the wave overtook them. It broke like a waterfall, sweeping them both overboard. Brutally shoved, *Njord*'s entire bow section dove and disappeared into the wave ahead. Lucien and Catherine hung on, she to the sheet, he to the tiller; they had both been thrown to port. Slowly, *Njord* surfaced again. All around, the sea seemed calm, as if the enormous wave had flattened it.

Seeing their boat, they could see themselves emerging from the abyss. During those last few seconds, they probably did not really have time to realize what was happening. They could not describe, in a word or a sentence, either their terror, or what had taken place; nonetheless, they had experienced both, through that direct knowledge of events shared by all those who have met and tried to face the unexpected. They had been those bodies buried by the wave, torn from the boat, thrust into the foaming water.

And now, hanging onto *Njord*'s hull, catching their breath, they waited while the devastating nightmare cataract sluiced off, leaving them exhausted, haggard, close to disbelief. What had happened? The wave, yes, that enormous wave. But did *Njord* actually disappear for a few moments? No matter. This was no time to try to re-create events that escaped them even as they tried to pin them down. Only one thing counted: getting back on board. Would they have the strength?

The constant motion of the boat did not make the operation any easier. Luckily, there was not too much freeboard aft. Just a few more seconds, to get a grip on themselves. Gasping, Lucien explained, "I'll give you a

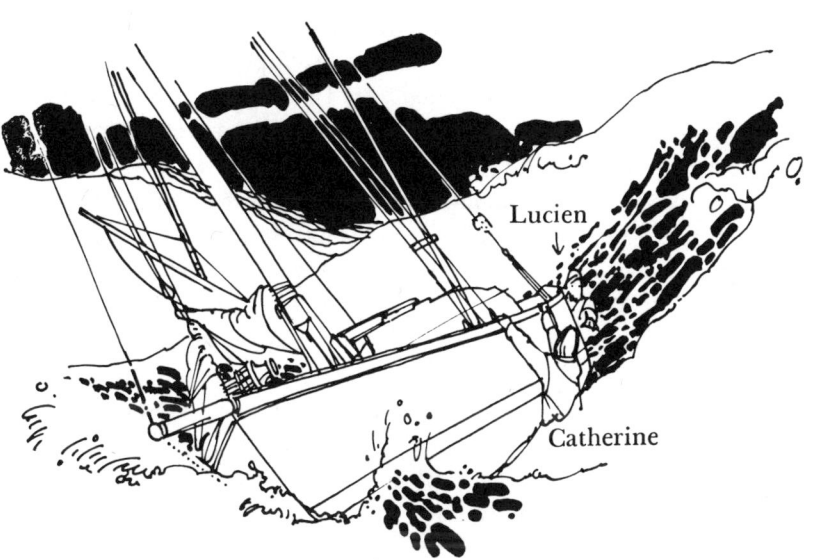

leg up. Make your move when the boat heels our way, then you can give me a hand."

There was no lifeline where they were. It may have been unwise not to have run it all the way to the stern —it might have saved them from being thrown overboard —but it made for one less obstacle to overcome in clambering aboard. The maneuver turned out to be hard, nevertheless. Catherine tried to climb the wall several times in vain, and was exhausted by each attempt. Summoning all her strength, she finally made it, found herself face down on deck, then slid into the cockpit. When she stood up, her legs were shaking with effort and relief.

Now to help Lucien. She bent down to him, grabbing his hand and oilskins. Pulling with all her might, suddenly he, too, was on deck, lying on his stomach. How did he make it? What did she do? Neither could say. They were aboard; that was all that mattered. What shape was the boat in, though? Lucien took the tiller, and moved it. As soon as it swung past dead center, it jammed to starboard, to port, it was obviously not acting on the rudder, which must have been crushed by the wave. It had been the sole wooden fitting below the waterline. Though the only mishap to be seen topside, it was a serious one. Forward, the two boats, dinghy and life raft, had not been carried away. Mast and rigging seemed intact.

There was complete chaos belowdecks. Every object, even the ones securely enough fastened to withstand the first capsize, had been sent flying, and usually broken. Unable to see, Catherine and Lucien had no way of figuring the damage, but as they felt around in the small cabin space, which they knew like the back of their hand, they realized that nothing was where it had been—not

Njord

even the solidly moored companionway. Everything was awash in the water that had poured into the boat. Lucien figured they had shipped some two tons of the liquid ballast.

In the face of this disaster, Catherine was heartbroken. She had worked so hard to stow things, to organize the cabin, and now it was wrecked . . . Now, even if the wind and sea quietened down, and a sunny day broke, the grim sight would still be there, with its ruined food, broken bottles, shattered partitions, its miasma of paper, food, clothing, paint, and oil, gradually soaking into the disgusting, greasy water that ruined everything it touched.

Meanwhile, *Njord* was slowly drifting broadside to the seas, rudderless, her poles bare. The feeling of calm that had followed the enormous wave's impact lingered on. The sea seemed quieter, though the wind had apparently not slackened. Was it more than just an impression? Perhaps, but the lull could be only temporary at best. What would happen if another breaking sea hit the boat? How would *Njord* respond to the new onslaught, overloaded as she was? Wouldn't she sink? There was the life raft, of course, but Lucien had bought it secondhand, and had not been able to have it tested before leaving. The raft was made in England, and none of the stores carried that brand. How recently had it been inspected? Two years, perhaps three? Would it work in an emergency? In their minds, numbed by fatigue and stress, an obsession began to grow: the life raft. "I was sure it wouldn't inflate when we tried to use it," Lucien later explained.

A moment later, he could not stand it any longer. The life raft was there in its bag, lashed to the cabin top aft of the mast. He was going to test it. He just did not think further. Why bother, since it would not inflate anyway? Slumped in the cockpit, dead tired, Catherine made no move to interfere.

Lucien climbed to the cabin top, seized the toggle, and pulled. The raft unfolded immediately, caught the wind at the same instant, and nearly blew overboard. Grabbing it, Lucien managed to get the raft under control, and wrenched it away from the shrouds, where it might have ripped. He dragged the raft to the cockpit, but there was no room. The large round basin measured nearly six feet across, and it was topped by an inflated arch-like tube supporting a nylon tent. The only solution was to put the raft in the water, solidly moored to *Njord*.

Right away, a new worry made itself felt. What if the line parted? Their last chance of survival would vanish. They had to put the raft in the water, of course, and tie it to the boat. But further, they had to get into it themselves, while waiting for the fury of the elements to spend itself. They neither trusted *Njord,* disabled as she was, nor had the heart to take shelter in the devastated cabin, which could be swamped by another wave at any time. They would feel safer in the life raft.

From then on, everything happened very fast. While Catherine struggled to hold the inflatable alongside, Lucien collected a few things, furiously sifting through the piles of debris, and yanking them from the chaos belowdecks: some tins and distress signals, two five-gallon jerry cans of water, a compass—miraculously intact—and, after a long search, the leather bag with their money and papers. This would allow them to eat, drink, guess at their position, signal to eventual rescuers, and manage, once ashore. Catherine clambered into the raft. Lucien handed the gear over, then joined her. It was 7:30 P.M.

"As soon as we were in," Catherine later reported,

7:50 P.M.

"I felt safe. Maybe we could just wait for things to get better . . . At the same time, though, I felt a wave of overwhelming anxiety, a confused feeling that we had taken a step that could not be undone."

The life raft, which tended to drift faster than *Njord,* stayed under her lee. Through the tent opening, they could see their boat's after port quarter, a few yards away. Suddenly, the mooring ring that the line was tied to ripped out.

"It was awful," said Catherine. "For a few seconds, we could see *Njord*'s black shape standing out like a ghost ship, and then nothing, because the raft turned. When it came back to the right position, *Njord* had disappeared. I felt a wave of pain through my whole body. Lucien was stricken. He wanted to jump overboard, to try to swim after his boat. I had a hard time convincing him not to.

"In the beginning, I had tried to hold the end of the line connecting us to *Njord* myself, but it was too hard, and I had to give up. Lucien then suggested that we tie it to the raft itself. I nearly said: 'It can't be done. It's pulling so hard, it will never hold.' But I kept quiet. I knew the mooring ring was going to tear out, but I also knew we were not strong enough to hold on by ourselves. We were just too tired. There was only one thing to do: tie the line on, and hope against hope that the ring would hold. I must have really believed it would, too, because when the ring let go, it just tore my heart out."

"For me," said Lucien "it was like having my boat and our future both disappear at once."

2. First Conversation: Heading for Disaster

"The die is cast now. You have little chance of ever seeing the *Njord* again. Here you are in your inflatable life raft at the mercy of a wind climbing past force eight and a sea gone mad. What is your position? As far as your course can be reconstructed, you are approximately 41° 20' latitude north by 5° 10' longitude east, which is to say about the latitude of Barcelona and the longitude of Marseille. Thus you are a hundred miles or so from the nearest French shore or the Balearic Islands, and roughly a hundred and twenty miles from the Spanish mainland.

"Corsica and Sardinia are farther away. If you could choose your course, you would have to cover one hundred and fifty miles to reach the nearest point of either island. But you have no choice. You are at the mercy of the wind that sweeps you toward southwestern Sardinia. And between you and southwestern Sardinia lie some two hundred and forty miles, virtually an insuperable distance.

"Of course, you can hope that the wind will veer into the northeast and take you toward Spain or the Balearics.

But for you to reach one of these coasts, the wind must blow from the same direction for at least five to ten days. All things considered, your greatest hope seems to lie in meeting a ship.

"Were you aware of this situation and, if so, how did you feel about it?"

Lucien: I was absolutely crushed at first. I didn't want to believe it. I told myself it wasn't possible our lovely adventure could end this way. And at the same time, I wasn't really worried. I was sure that we'd be rescued by a boat quite soon.

Catherine: I wasn't at all sure of that.

Lucien: Yet you kept quiet about it.

Catherine: No. But from the very first I knew it was a question of survival. I made a mental inventory of what we had on board. I decided that by being careful we had enough to last a fortnight. If we didn't reach land or weren't rescued by then, we would be done for. For the moment, we were exhausted and I felt the most important thing was for us to try to sleep. But there was about an inch of water in the bottom of the life raft and, although we were soaked, we didn't feel like lying down in it.

Lucien had lost his boots when we were thrown into the sea, but I still had mine. I took them off and we tried to bail with them but without much success. We gave up finally and sat facing each other on the jerry cans we had brought with us. Harassed and empty-headed, we began to doze.

Lucien: At first, we spoke of the future, what we would do after we'd been rescued. That shows we were not really worried. In fact, one of the things that bothered us the most was the thought of losing face in the eyes of our friends ashore. We'd left on such a long, terribly long journey . . . and here we were after only three days. And the boat we'd abandoned . . . I really loved her. It wasn't possible. We would find her again.

Catherine: He was speaking but I was already asleep despite the discomfort, the cold, the noise of the wind and the sea, the tossing of the life raft. His voice no longer reached me and after a while he stopped speaking.

Lucien: Yes, I'd fallen asleep, too.

"You were able to rest a while that way?"

Lucien: At least three hours. We'd left the *Njord* at 8:00 P.M. We woke up, or rather regained consciousness, a bit before midnight. I think it was the increasingly violent movements of the life raft that snapped us out of our stupor, plus the fact that we had recovered a bit. Incidentally, it was a miracle that we had the benefit of those few hours of respite. Had we not had some rest, I doubt we could have faced what lay in store for us.

"When you returned to reality after this rest, what struck you the hardest?"

Lucien: First, that the raft was deflating. The raft consisted of two inflatable rings, one on top of the other, surmounted by an inflatable arch that supported the tent. The inflatable arch was collapsing and the upper ring was softening. There were two valves, one for the lower ring and the other for the upper ring and the arch. In one of the built-in pockets there was a pump for reinflating the raft. But how could we use it in the dark? We knew only two things: the life raft was deflating and, thanks to our groping, we had found a pump. By using it, we ran the risk of deflating the raft entirely while handling unfamiliar valves. That was our biggest worry at the moment.

At one point Catherine found something at the top of the arch that could have been a valve. Judging from its position, this valve should have affected the inflation of the arch alone. So we didn't run a great risk by trying it. The main thing was to conserve the air that filled the two horizontal rings. We tried twenty times to attach the pump to this piece of metal, each time in vain.

Later we realized that what we had taken to be a valve was an electric socket. Originally, the raft must have had an electrical lighting system, no doubt battery-powered. It consisted of one bulb on the inside and two running lights, the sort no passing boat would see—short of a miracle—and that would stop working after a dozen hours.

But at that moment we were in no position to philosophize on the structure of the raft. It was deflating and that was alarming enough to be our main concern. At the same time, we were aware that the wind was increasing and the seas were building. The raft was being tossed about more and more brutally and unpredictably, while the din grew louder and louder. From within the tent, it was impossible to see what was going on, but we had the impression we were surrounded by cresting waves. When a big one hit the raft, we probably submerged completely. At least that's how it felt.

Catherine: Yes. I remember we said to each other, "That one must have finished off the *Njord*," which was absurd. There was little chance of the same wave threatening the *Njord* and then us. In a way, it was reassuring that the raft behaved well. But, even assuming it didn't deflate too quickly, how long could it stand up to such unleashed fury? The fact that we had begun to ask ourselves that kind of question gives some idea of how worried we were. We felt like people hurtling down a dizzying slope in some uncontrollable vehicle, not knowing when it would disintegrate or leave its course to crash against a tree or overturn in a ravine.

With every buffet, with each more sickening lurch I felt my muscles contract, and I sensed the same thing was happening to Lucien. I wasn't hungry anymore, I wasn't cold. I was afraid. It was physical fear—welling uncontrollably—not panic. It was the agonizing wait for something to happen, something decisive, irreversible. I didn't know what, exactly. I suppose that's why I wasn't surprised when I felt the raft move more abruptly than before and in a different way, as though this movement helped to hoist it to the summit of something. Was this the end? I scarcely had time to form the question before we capsized.

Lucien: Fortunately, I had taken the precaution of installing myself next to the tent flap which was fastened by snaps. I was able to jerk it open immediately and we landed in the water again and clung, panting and gasping, to the lifeline around the raft. The whole thing lasted only a few seconds.

The water was warm, warmer than the air. I was terribly worried. "If it goes on like this . . ." I fretted. How could one possibly survive, even a few days, in a life raft that threatened to capsize at any moment? Until then, we had relied on its stability . . . And would we even manage to right it? I asked Catherine, "Are you all right? Do you still have the strength to make an effort?" She replied, "Yes, I'll have to. We're not going to die here."

Catherine: It's true. Suddenly I had all my strength back. Now that the thing I had dreaded had happened, I was no longer afraid. We were still alive and the raft was still intact. All we had to do was right it. I was sure we could do it. But somehow, I was a bit disappointed. I naively imagined that these inflatable life rafts righted themselves automatically if they capsized.

We caught our breath for a while. After all, we weren't badly off in the water. It was warm and the sea

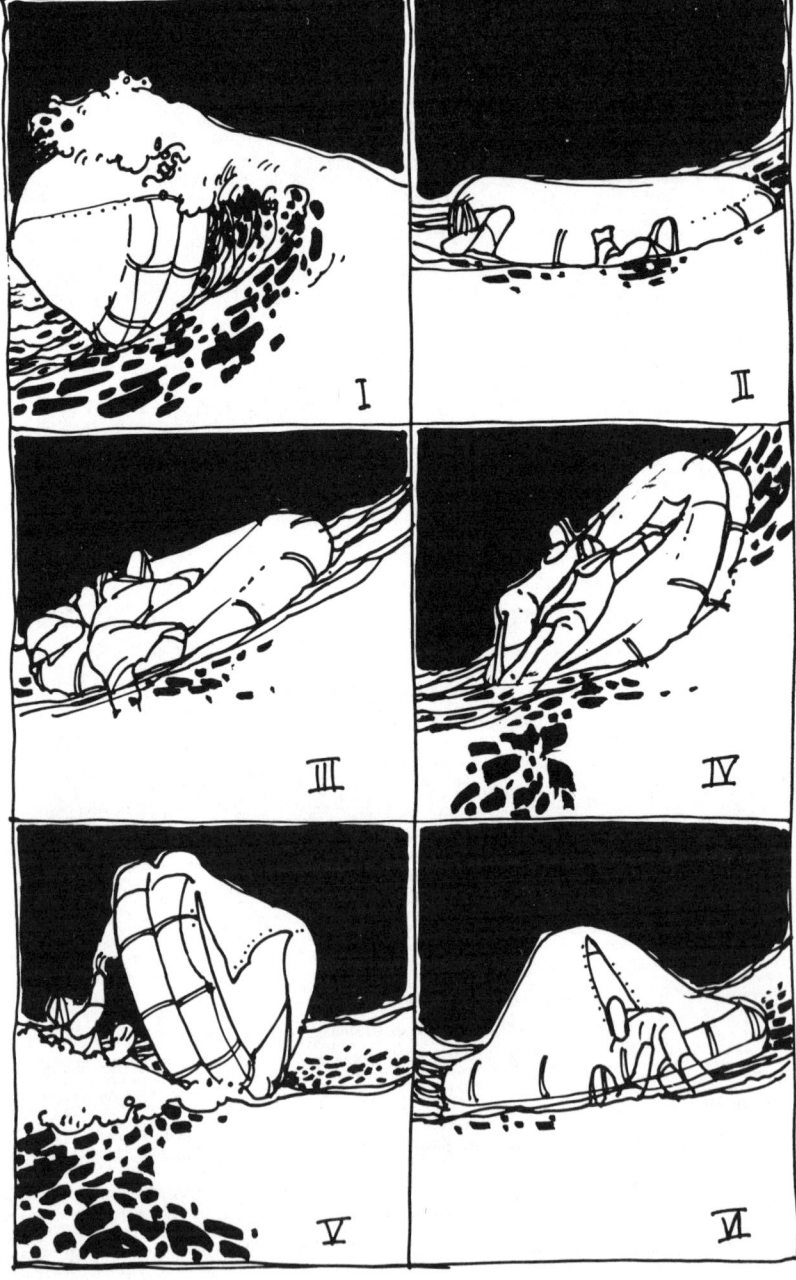

seemed less cruel from there. One could see only a few meters in any direction.

Lucien: I was wondering what was the best way to right that wretched tub. We began by hauling ourselves up onto the bottom, using the lifeline as a ladder and the antidrift pockets as something handy to grip. Then we lay flat on our stomachs and caught our breath again. Finally, resting our feet against the lifeline, we threw our weight backward while pulling on the antidrift pockets. To do this, we had to wait for the right moment when the raft, climbing the flanks of a wave, was already tilted almost 45 degrees. We succeeded on the first try. Now all we had to do was climb back aboard which, even without a ladder, posed no real problem.

Catherine: No. But our problems began once we were back inside our ridiculous rubber basin. Almost everything we had taken with us, everything on which our survival depended, had disappeared during the capsize. Even my boots that I'd not put back on after we'd tried bailing with them.

It was hardly a big job to inventory what was left: one flare, three sticks of phosphorous, one tin of corned beef that Lucien had put in one of the pockets of his oilskin, and, fortunately, one of the two jerry cans of fresh water. But the jerry was only half full. It must have leaked when we'd sat on it. Perhaps it was badly sealed. In any case, we had only ten liters of water. And in a way it was a miracle that we had any left. How had the can stayed in the tent? Why hadn't it gone straight to the bottom like the other one? Perhaps the very fact that it was half empty, the air trapped inside, had allowed it to float. What if it had been full . . . ? I still shudder at the thought.

Another thing. The pump was still there. We'd put it

back in its place after our first attempts to pump up the raft by means of the electric socket. I don't know if we realized at the moment just how important it was not to have lost it, but I know now that its loss would have cut our chances of survival almost to nothing.

Oh yes! We had one more thing, something utterly absurd under the circumstances. The leather pouch containing our money and identification papers. We had slipped it into one of the pockets of the raft.

"What was your reaction then? Resignation, despair?"

Catherine: Neither, as far as I was concerned. Not having anything to eat didn't frighten me. I knew one could last about twenty days without food, and we had water. What frightened me was the cold, especially in view of how tired I was. I was afraid of falling into a sort of torpor and being trapped inside the tent and drowning if we capsized again.

"Hadn't you thought of removing the tent?"

Catherine: No, it protected us from the wind and the waves, and thus from the cold. Besides, it wasn't detachable. Removing it would have meant cutting or tearing it, and we certainly didn't want to risk damaging the raft.

Lucien: It wasn't the cold that frightened me. Besides, I wasn't really cold. What really worried me was the loss of our provisions. I thought, "Without food we'll soon lose our strength. We won't be able to continue our struggle for survival."

I was equally upset about losing the flares. Without some means of signaling, no boat would see us at night. Of course, we might sight one during the day. But even so, our chances of being seen were rather slim. Yet, at the same time, I thought our one remaining flare might be enough to bring about our rescue. One had to cling to something.

"What time was it then?"

Catherine: Midnight. We had abandoned the *Njord* about four hours before.

"What did you do once you were inside the raft again?"

Catherine: We began by making everything secure. We couldn't rule out the possibility of capsizing again so long as the sea was so rough. Next, we settled close to each other to try to warm up a bit and to doze. But since we had only one jerry can left, we dared not sit on it for fear it might leak again. So we half lay in the bottom, that's to say in the water, with our legs drawn up and our backs resting against the inflated gunwales, while the raft was tossed about at the whim of the waves and our ears rang with the roar of the wind and the boom of the sea.

The arch had continued to collapse, greatly reducing our room. The tent canvas sagged in our faces, which only heightened our anxiety. We had tried one more time to use the piece of metal we thought was a valve, and then had given up. How much longer would this raft hold enough air to support us? But once more fatigue was stronger than fear and we soon sank again into a kind of semiconscious state.

Lucien: We slept and we talked by turns. We were trying to evaluate our chances of survival. What should we do? What would happen to us? Or then we would ask each other, "How do you feel? You're not too uncomfortable? Are you very cold?" Then suddenly, about 2:00 A.M., we capsized again. I don't know whether it was because we were dozing that we didn't feel it coming but that's how it happened. It was so violent we didn't have time to react. The first time, there had been a warning of sorts, time to think, "Something is going to happen." That had been enough to alert us. As soon as it had become obvious that we were going to capsize, I'd opened the tent and we were out. Whereas the second time, before we knew what had happened, we were in the drink, pinned inside the closed tent.

I began thrashing about like a madman trying to find the opening, but I couldn't. Suddenly, death was real. I thought, "This is it, you're going to die." How long did it take to find that damned opening? I don't know, perhaps fifteen seconds. But it was a long time, a very long time. And when I reached the surface, after having found the opening, I realized I was alone. "Catherine," I yelled. No answer. The first time, the moment we'd capsized we'd shouted, "Catherine, Lucien, are you all right?" and we'd been reassured right away. But this time there was no reply. I raised the raft and put my arm inside. Catherine was there. I pulled her by her oilskin and she surfaced, alive.

"And how did things go for you, Catherine?"

Catherine: The capsizing also took me by surprise. I didn't have time to aim for the exit and I got caught in the canvas. I struggled, but in vain. I, too, thought I was going to drown. Yet, I didn't swallow any water.

As a matter of fact, I think I panicked a little because there must have been some air trapped inside the overturned raft and my life jacket should have kept my head above water. Today, I wonder whether I didn't continue breathing without being aware of it. But I was still caught in there, not knowing what to do. I wasn't even aware of Lucien pulling me out. I found myself on the outside again without understanding what was happening to me.

Lucien: This time it took quite a while to catch our breath. I was cursing. "Saving our precious little lives isn't going to be easy," I thought. We were clutching the lifeline of the raft and thundering away at the thing. "What kind of contraption is this? It's worthless." Then we tried righting it again as we had done before, but our brief struggle against drowning must have exhausted us because we had a lot of trouble doing it.

Just hauling ourselves, stomach-down, onto the raft was quite an undertaking. We had the feeling we'd never make it. But it's usually at that moment, when one is nearing the end of one's strength, that one finds the energy for one more try. How did we go about it? I don't know. We were just back on the raft, side by side, utterly exhausted. My tongue was hanging out of my mouth. I heard Catherine panting. We stayed that way a long time, unable to move. How long? There's no way of telling. Ten minutes, perhaps much less. In such moments, time plays odd tricks. And then we had to right the raft, which wasn't easy either. We managed only after several attempts. Our timing was off. We were always one beat behind the waves, so that all our efforts resulted only in tilting the raft slightly before it fell back into the trough.

Problems also arose boarding the raft again. The arch was so badly deflated that we had to crawl in order to slide under the tent. No doubt that was why we had had so much trouble finding the way out. If the tent had been taut, we might have escaped more easily.

Catherine: And it was at that point, when we had just settled ourselves on board again, that you glanced outside and saw the lights, the lights of a big boat.

Lucien: Oh yes! I said to you, "There's a boat, we're saved."

3. As Far as One Can Go

The flare, the only flare. Lucien's first move had been to take the flare from the pocket where he had carefully stowed it. The flare was a cylinder about thirty centimeters long and three to four centimeters in diameter, a red flare that had to climb into the sky, a flare that had to alert men who were amusing themselves aboard that ship during what was for them an insignificant storm, letting them know that a few meters away from their safety others in dire peril awaited their help.

Lucien looked again toward the lights that had appeared suddenly in the heart of that night gone mad, in that deserted sea. No, it was no mirage, no figment born of fatigue. There really was a ship passing within less than a mile of them. It was apparently coming from the northeast and heading southwest. Having departed from Genoa or Nice it was no doubt making for Gibraltar or Tangiers. Lucien could see perfectly the red light on its port side.

Catherine was on her knees in the bottom of the raft, for it was impossible to stand by holding on to the arch, which had become almost totally flaccid. She, too, saw the red light approaching, followed by the vague glimmerings of lighted lamps in some of the cabins—snug places, warm, dry, and safe.

As Lucien set about igniting the flare, Catherine could not overcome her anxiety. One of her friends had blown off his hand while launching the same kind of flare. For a moment she imagined such a mutilation—Lucien collapsed in the middle of the raft, his blood spurting over her.

He crouched, rested his left forearm against the inflatable gunwale, and held the base of the flare toward the sea. He pulled the cord that worked the detonator. A short white flame leapt forth that lighted their faces a second, while a red ball rose with a hiss like that of skis crossing snow.

Lucien emitted a groan, released the charred tube and took his left hand in his right. "Did you hurt yourself?" Catherine asked.

"No, it's nothing, a slight burn."

The red ball proceeded along its aerial journey, but it had been launched at a low angle and downwind so that it attained neither maximum height nor distance before it flared fully, before it became that bizarre flower that would begin its slow swinging descent.

That incandescent red beacon, now hanging from its parachute, was carried away by the wind, but it lighted up the sea, the waves, made the life raft blaze. It would have been impossible not to have seen it. The sudden flaring of that meteor could not have gone unnoticed. Lucien had forgotten about his burn. Catherine and he had become nothing more than a question addressed to that ship, plying its course, toward which the flare cast its

appeal. Had someone on the bridge seen the signal, sounded the alarm? Was this disastrous voyage nearing its end? Scarcely had they begun to hope when they felt themselves thrown back into the heart of the gale. Nothing was happening. The flare had already become an imperceptible dot of light descending toward the sea and there was still no indication that it had achieved its purpose.

And yet, yes! The ship seemed to slow down. Soon it became clear that not only had it cut back its speed but it had altered its course. Before, only its red portside light had been visible. Now, the ship's bow was pointing at the life raft. Its green starboard light came into view. Then the red light disappeared. The ship had made almost a complete semicircle.

To indicate his position more precisely, Lucien ignited one of the three sticks of phosphorous. He burned himself again, more seriously than with the flare. No matter. All his former strength had returned, especially when a searchlight came on and began to sweep across the sea. "We're saved," he shouted. "I tell you, we're saved."

Catherine didn't reply. Once the first moment of exaltation had passed, she sought refuge in her customarily prudent attitude. Saved? That was taking a lot for granted. Would the ship ever manage to find their tiny rubber raft on that dark night amidst those troughs? And even if it did, it could lose them again. Besides, could it bring off such a delicate maneuver in the middle of the night, in such violent seas? Was it possible for this mass of several thousand tons to draw alongside the raft without crushing it, to hoist the shipwrecked pair on board without dropping them into the sea or dashing them against its steel side?

She had not lost all hope, but she was worried about what might happen and at the same time afraid to give in to an unreasonable enthusiasm. Should it prove unjustified, it could destroy her remaining strength. But at the same time she followed intently the movements of the searchlight that painted a brilliant elliptical patch on the water, back and forth, while the ship slowly approached them.

Minutes went by while the puddle of light failed to reach them. To guide the searchlight, Lucien ignited the last two sticks of phosphorous, burning himself again each time. The beam of light moved to the right as if their would-be rescuers, tired of not finding them in the sector where in fact they drifted, had decided to seek them elsewhere. At the same time, the ship seemed to have stopped. Blown by the wind and carried by the waves, it drifted more rapidly than the raft, and the distance between them slowly grew.

The search had begun almost two hours ago, and now seemed hopeless. It would have taken another flare to show their position again, for it had become increasingly clear that no one on board the big ship had the slightest idea where to look for them. Furthermore, it looked as though, for the moment at least, the hopeless search had been abandoned. The searchlight had gone out, perhaps in anticipation of another signal. The ship had not resumed its course. Its stationary green running light continued to pierce the night.

"They're waiting for daylight," Lucien said. "They'll find us at dawn."

They sat on either side of the tent opening and stared dully at that green light that grew ever more distant. Fatigue, banished for a while by the excitement of the prospect of being saved, reasserted itself. Yet the chance to rest was to be denied them. At the very moment when they had decided there was nothing left for them to do but await the dawn and try to doze, a wave, more abrupt than the others, capsized them for the third time.

They were taken by surprise. Throughout their entire vigil of the ship's progress, they had scarcely paid any attention at all to the tossing of their raft whose movements had become part of their universe. It was undoubtedly their preoccupation with the ship that led them to believe the wind had died down a little and the sea had become less violent. Yet despite their surprise, they had no trouble getting out of the tent this time; they had been virtually outside it with their heads already in the opening when they capsized. "This is hardly the time to drown," growled Lucien.

While in the water, clinging to the overturned raft, they could see, whenever they reached the top of a wave, the green eye of the big ship watching over them. Perhaps, in a few hours, they would be on board. How sweet it would be to undress, to wash away the salt and to sleep, to sleep for hours in a dry bed, safe in the hollow of a stable world, without the perceptual threat of being engulfed by seas.

They would go to Gibraltar. There they would issue an alert. If *Njord* had not gone down, if she had managed to elude breaking seas, perhaps she would be found floating nonchalantly on a quietened sea. Storms did not last forever. How long had this one been raging? They no longer had a clear notion of time. So many events had filled those past hours that they had seemed to burst, scatter, and multiply. They had left Porquerolles, and gone in search of islands, a profusion of islands, of sun, of friends, of laughter, of life itself! They had left Wednesday morning and it was not yet dawn Saturday. Three days, scarcely three days. The storm had begun in earnest yesterday morning, just twenty-four hours ago. And it was only eight hours since abandoning *Njord*. Eight hours . . . and already they had capsized three times and had lost almost all they had brought with them for survival.

Lucien and Catherine were relying on their life jackets to keep them afloat, clinging to the lifelines of their overturned raft while the ship, several hundred meters away, seemed to ignore them.

They were putting off the moment when they would have to make the effort to right the raft. They were comfortable in the water. They had let themselves go and were relaxed. A marvelous numbness had set in. "If only a wave would knock it over again and put it right side up," thought Lucien. But they were not to be so fortunate. Upside down, with its flat bottom up and the tent acting as a keel, the raft seemed to have found its most stable position.

Lucien closed his eyes for an instant, falling halfway into the insidious snare of sleep, but then roused himself abruptly. "This is hardly the time to drown," he murmured again. Catherine had already been able to hoist herself into the raft. He rejoined her. Together they managed to right the raft on their first try. Several seconds later, they were once again under the tent that sagged around the collapsed arch. The upper inflatable ring was now halfway deflated. Fortunately the lower ring still held firm.

"As soon as there's light we'll find a way to pump up this bloody tub," said Lucien while glancing through the tent opening to make sure the green light was still there. He thought, "If I still had the flares I could attract the attention of the crew again, convince them once and for all that there really are people in distress here. What will they think if we give no further sign of life? That we have finally disappeared or that whoever saw the signal was the victim of some mirage? If only our signal hadn't been seen just by a simple helmsman . . ."

But his optimism returned: "They spotted us, looked for us with their searchlight. They know we're here, somewhere in the midst of this raging sea. They won't leave before having found us."

Catherine was dozing. Such hope as she had left was succumbing to fatigue. Not that she despaired, but she had tried to convince herself that she no longer believed in the ship, in the rescue. In her mind she saw the raft beach finally on some distant shore. "I always thought," she said later, "we'd have to drift a long time in that raft, that we'd have to hang on for days and nights, hoarding our strength, before we got anywhere. All I asked was that the wind die down, that the sea grow calm, that the day come and with it the sun and warmth. Then everything would be easier. I wasn't thinking about death. At nineteen, it's hard to think about one's own death."

Meanwhile, little by little, the ship had been drifting farther away. It was now nearly two miles distant, whereas at one point it had been as close as 800 meters. The sky was growing light in the east; it would be daylight in less than an hour. This time the sun would be visible, for one could make out several stars. The wind had finally got the upper hand over the mass of clouds whose sooty reflections had made the sea even more hostile the day before. Soon they would be free of that cruel night.

Day began to break. They could no longer make out the green eye that had watched over them and that could have been mistaken for a decoy. They could now see the ship itself, a large freighter with an after bridge house and derricks springing from its decks. It began to move, to make a semicircle. It could only mean the beginning of a renewed search. Those sailors were going to cross rule the sea, to rake it, to find those who sought their help. No. They were giving up. The ship, after having stood by for four hours, was resuming its course, was heading southeast. A few moments later, it was nothing but a scratch on the horizon while the icy wind, saluting the sun's arrival, blew stronger than ever.

Lucien was broken. Catherine tried to comfort him. "Those people wouldn't have left unless they'd sent word for help. They've entrusted the search to others better equipped than they. A few hours from now planes and helicopters will step in. They'll find us, they'll save us. We just have to hang on a bit longer."

And then it was light. It was light for the first time since they had abandoned *Njord*. For the first time they could really examine the rubber raft, look for the location of the valves, see how they worked, and attach the pump and reinflate the arch and upper ring. It made them smile to discover the electric socket over which they had slaved for so long.

Next, they discovered that the pump could also be used to empty the water that had accumulated in the bottom. A dry boat and a well stretched tent, which gave them the impression of greatly increased room, improved their living conditions, although outside, the condition of the sea and the strength of wind were as alarming as ever. The inventory of resources offered by the raft proved disappointing. Aside from a sea anchor, a ball of nylon cord, and a small empty canvas bag, they found nothing. On boarding the raft the day before they had noticed the presence of a pair of oars but these had disappeared, no doubt lost during the first capsize.

There was daylight now. The world was no longer that terrible arena within which they had struggled for hours—a basin, a sort of spherical aquarium into which they had been hurled. The waves had once again become real waves and not obscure forces.

The wind still howled, of course, and the breakers still boomed, holding the raft in their terrifying run, but the sun warmed a little through the canvas. The water trapped under their oilskins, soaking their clothing, rose steadily to the temperature of their bodies. A mouthful of corned beef taken from their only tin, a few swallows of water, and hope reappeared briefly, only to be shattered by the brutal impact of a wave of more than average violence.

Catherine was estimating their chances of survival. There was enough fresh water for four days. And in September there was always the possibility of rain. If only the wind would drop, if only they could catch their breath a bit. As for food, well, never mind, they would do without. If it rained, they could last another ten days perhaps, providing also they had sun and were able to undress and warm themselves. But what was the way out? A ship? Catherine had less and less faith in that. By day a ship would have to run across them. Moreover, it was quite possible a ship would ram them without seeing them. At night, without flares, they didn't have a chance.

As long as this storm lasted, there wasn't the slightest hope of rescue. Hundreds of miles of empty sea to be crossed, perhaps as far as Africa. And always the danger of capsizing again and finally not having the strength to right the raft. If only they could sleep . . . But despite fatigue, their minds remained alert. Each shock felt aboard the raft could mean the onset of capsize. A few seconds of dozing, a few seconds of tension, a moment of warmth, and then sudden fear that turned them to ice. And perhaps at that moment a ship was passing by and they couldn't even see it. One of them should have kept a constant watch, on the outside, but that wasn't possible. They no longer had the strength. They had to sleep.

That was the way things were when they were startled by a noise that drowned out everything else. "The noise of a train," Catherine said later. Lucien said, "I thought a huge ship was bearing down on us. I was afraid." For several seconds they were petrified, listening with their whole bodies to the growing roar. They faced each other without a word, kneeling in the bottom of the raft, their hands clenching the inflated rubber ring.

And then came the impact. The world exploded about them. They were in the water, swimming in a cauldron of boiling foam, whipped along by the breaker's turbulence. "I immediately thought the raft had disintegrated," Lucien said later, "that it was all over, that we were nothing more than two orange specks lost at sea, bits of flotsam drifting forever." Yet no sooner had he gone through that new death than he found himself on the surface again, swimming in a leveled-out sea, flattened by the enormous roller whose passage had marked the water with a wrinkled network of watery scars.

A few arm lengths away, he saw Catherine emerge from her brutal burial and, not far from them, less than fifteen meters to windward, the overturned raft, bobbing like a cork. Fifteen meters, under such conditions, was the distance separating life from death—a tiny space, but an eternity. Crossing that trifling yet gigantic gulf, dragged down by oilskins and clothing, seemed an insurmountable task. Yet they did it, stroke by stroke, afraid until the very end that their strength would fail them before they covered the last few centimeters.

Once again, they tried hard to catch their breath while hanging on to the raft. When they had last capsized during the night, they had already been on the point of giving up, of surrendering to their fatigue. Yet, then, there had been the ship whose watchful nearness had given them reason to hope and therefore to fight. There was no such incentive now. The storm, far from having abated, seemed unable to reach a climax. Their exhaustion continued to mount. It reached a level beyond which death seemed a certainty. Yet, in what they considered their ultimate effort (Yes, thought Lucien, it was doubtless the last time; after that they wouldn't be able to do it again), they managed to haul themselves onto the bottom of the overturned raft. But that wasn't enough, they had to go further, again had to extend the limits of possibility, limits they kept thinking they had already reached. They had to seize the antidrift pockets, arch their bodies and snap them so as to right the raft.

Later, Lucien said, "Let's be frank. It took quite an effort to right the raft. I don't think a man can do it alone, unless he's in top form, which must seldom be the case for a shipwrecked person. Even with two people, if we didn't make the effort at the precise moment it was needed—just before reaching the crest of a wave—and if our movements were not perfectly coordinated, our efforts were wasted.

All we could do then was wait, limp and helpless, until we found the energy for another try, energy which seemed less and less likely as time passed.

"At the time of our first capsize about midnight, we had been able to do the necessary maneuvers rather easily despite the darkness and our lack of experience. And we had thought then we were on the edge of exhaustion, which was utterly absurd given what happened later.

"The next two times, it had been a bit more difficult. Yet it never really occurred to us that we couldn't manage it. But this time, after having been rolled over by the breaker, having had to swim to reach the raft—which we'd not had to do the other times—after all we'd gone through, we really began to wonder. Even when we were lying on the raft bottom, we wondered if we could right it. We must have tried it several times. I don't know how many, perhaps four or five. More times certainly than after we had been trapped in the tent and thought we would drown. Worst of all, each attempt was more painful and, I began to realize, less effective. We had to wait longer to find the strength for another try. We managed at last, but only to discover that the wave had torn away our tent. Nothing was left but shreds, hanging from the arch."

They were completely crushed by their discovery. At once they realized that they must survive from then on without the slightest protection, exposed to the wind, defenseless against the cold, constantly soaked by a raging sea hurling its foaming crests in every direction.

The way the raft had been before now seemed to them like a comfortable lair, a closed universe in which they could take shelter. From now on, they would be irretrievably condemned to the out-of-doors, exposed to all the elements. "This time, it's all over," Lucien murmured. Catherine, despite her ardent desire to live, was not far from the same thought. They had reached the point where they were wondering whether it was really worth making the taxing effort of climbing aboard the upright raft again.

Yet at the same time, something within them drew back from surrender. Catherine was the first to pull herself together. She put her hands around the arch. Her muscles tightened as she tried to snatch herself from the sea, but she slipped back. "I can't make it," she said. Lucien went to her aid. He too gripped the arch and prepared to give her a boost up, just as the evening before when it had been a matter of climbing back aboard *Njord*. With one push, he propelled Catherine onto the rubber gunwale. She lay there face down. With another effort by Lucien, she was safely in the raft, at least for the moment.

She caught her breath and then gave her hand to Lucien who also managed to climb back aboard. Once more they sat facing each other while their skiff raced from crest to crest, threatened at every moment with being submerged or overturned. A wave exploded next to them, showering them with spray and leaving a good ten liters in the life raft. They did not react. They remained awash in water that rose every time a sea slapped them in the face. And then their despair transformed itself into fury. "It can't last," Lucien shouted. "We must do something. We can't just die."

Do what? Try to bail, of course, to keep the raft from filling up. But the raft was too full to pump out. The pump was useful only for mopping up.

Lucien thought of the leather pouch in which they had stowed their money and papers. After having put those

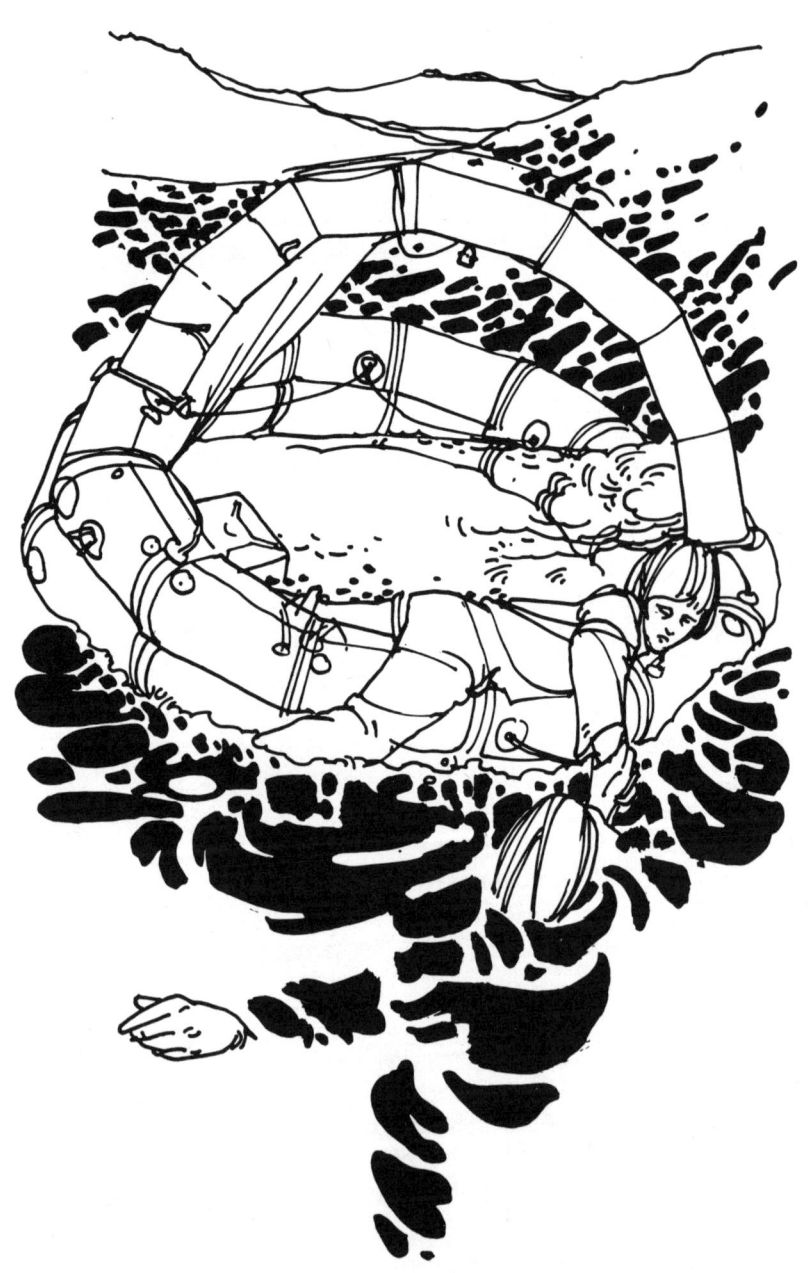

laughable mementos into the canvas pocket that was part of the raft's equipment, they tried to use the leather pouch as a pail, but without much success. Scooping water out with their hands was more effective, but a breaking wave dumped more in a second than they had been able to bail out. Besides, they were too tired to continue. Disheartened by the futility of their efforts, they gave up and sat on the rubber ring, one at each end of the arch to which they clung.

The sun was already high. It was nearly 10:00 A.M. Nothing hid from them any longer the spectacle of the wind-whipped sea, a sea that undulated, burrowed, twisted, and spewed up gluts of foam. In spite of their anguish and terror, they could not help being enthralled by such wild and demented beauty. "An unforgettable sight," they said later. It was impossible to resist its spell.

Now that they could see the waves coming, they began to realize that, while sitting on the inflatable ring, they could balance the raft to reduce the risk of capsizing. All they had to do was shift their weight in the right direction at the right time. The principle was simple. The raft revolved slowly, at the rate of about one revolution every ten minutes. Thus under ideal conditions, one sat facing the oncoming waves, and the other sat with his back to them. The person who was downwind, who saw the waves approach, had to warn the other and lean forward a little to counterbalance the weight of his own body, while the other person prepared to lean back for the same purpose.

Sometimes, however, the fateful moment found them broadside to the wave. They then had to move to a fore and aft position, balance the raft, and quickly return to their former place to hang on to the arch. This drill, which they had to perform two or three times a minute, exhausted them and was only effective when there was not too much water in the raft.

They continued to ship water at an alarming rate but had given up bailing as ineffective. The raft filled little by little, inexorably. When at the end of two or three hours it was three-quarters full, its freeboard reduced practically to nothing, capsize became inevitable. The only consolation was the spilling out of water. They experienced a capsize every two or three hours during the day of the 16th.

Now that they were squarely confronted by such a fate, they knew how it would be. They could play an active part, however exhausting, in preventing too frequent catastrophies, now that they could foresee the moment when they would have to right the raft and climb back on board, and the situation took on a new light. In any case, they had no choice. Either they embraced the rhythm or they gave up, and that would be the end. "Do you think we'll make it?" asked Lucien. "Yes," replied Catherine. "I don't know why, but I have the feeling we'll make it."

The raft filled up, wave after wave. When the water reached their knees, the moment was at hand. When they capsized, they held on in such a way that they landed on the upturned bottom of the raft without really falling into the sea. They righted the raft on the first try—they had found the rhythm—then climbed aboard one after the other, helping each other in order to avoid unnecessary effort. Lightened, the raft regained speed. Once more wind and sea swept them along their desperate course.

What if the bottom ripped? The bottom ripping or

the inflated tubes bursting were their chief worries. The water shipped aboard did not weigh on the inflatable partitions, of course, but their liquid cargo was brutally shaken, subjecting the partition to tremendous pressure. They hoped the fabric would last.

There was no question of sleeping or even of dozing, yet they managed to arrive at a state of semiconsciousness that acted as a substitute for sleep. Their movements became automatic. At times, Catherine was able to forget her shipwrecked state. She imagined herself on holiday, playing in the surf on some deserted beach. At times they spoke of the future. "When we get out of this," Lucien explained, "we'll have had enough of the sea for a long time. Perhaps the best thing then would be to buy a minibus, fit it out so we could live in it and go see Africa. I've always wanted to go to Africa."

Catherine was not tempted by that project. She knew Africa, had lived there with her father. She had no desire to return. She dreamt of a sedentary life, of buying some land so rocky it would be cheap, as no one else would want to live on it. Arid, parched land, scorched by the sun, lizards that ran along walls scarred by time, tufts of wild and fragrant plants, and a huge room, cool and dark, to protect her from the heat.

She thought of one of her aunts with whom she sometimes spent her holidays as a child, a woman who must now have been in her fifties. She had always lived on a big estate that she looked after herself. "Perhaps I would stay with her a while, until I decide what to do," Catherine thought.

"Look out," said Lucien, "here comes a big one." With a movement of her back, Catherine balanced the rubber raft so that it slipped easily over the dangerous crest without taking on a drop of water. They enjoyed the satisfaction of having put off the next capsize by several minutes.

The sea had become completely white, and the larger masses of foam gave the impression of ships bearing down on them. They saw many motorboats and sailboats take shape from the waves . . . Nearsighted Lucien, who had long since lost his glasses, pointed them out to Catherine. "There! It's a boat. I tell you it *is* a boat." Catherine stared hard. Although mistrustful, she too let herself be taken in. Perhaps, perhaps it really was one. Their hearts beating hard, they followed that fabulous mirage, with their eyes, until the wind transformed it into a shower of spray.

Twenty times the illusion was reborn and twenty times they fell victim to it. The day drew to a close, intensifying the ghostly cycle, restoring cold and fear. The

few remaining grams of corned beef and a swallow or two of water would help them last until dawn. Lucien gashed the palm of his hand with a clumsy movement of the knife, but he felt nothing and did not bleed.

Just before dark, a ship passed within less than a mile. They shouted to attract the attention of its crew, but without success. They did not stop shouting until they had run out of strength.

4. Second Conversation: To the End of the Nightmare

"What was your situation at the beginning of the second night?"

Catherine: Very bad. It seemed to us the wind was as strong as ever and the seas even bigger than before. We were better off in one way, though, as the swell was more regular. We no longer had to worry about those damned cross seas.

At sunset, our liquid universe seemed to be animated by a very deep, regular breathing. Perhaps the wind had fallen. We couldn't tell, because the steady unrolling of those great sparkling valleys was broken from time to time by the explosion of a breaker which completely disrupted the order of our world without warning. Fortunately, we never found ourselves in the path of one of those breakers, some of which were really enormous.

We kept a sharp eye on them. Sometimes they broke so close to us that the noise made us turn around, and we froze in terror at the sight. We wondered what would happen to us if we were caught up in such a turmoil. We were afraid it would happen at any moment. We kept thinking that we saw it building up behind us: that wave a bit higher, a bit more translucent, with its crest a bit more feathery, a bit closer to losing its balance, that might topple over on us and engulf us. But no, those threats receded into the distance, one after another. When a wave came at us, the raft would take off, climb to the top of that moving green wall—that was the most terrible moment of all, especially when the crest was just beginning to curl—and then drop into the trough.

Sometimes a wave that had just borne us aloft broke the moment after we had crested it. Then we would see its back, marked by a thousand transverse ridges, curl over and collapse in a frenzy of noise and foam. And we would say, "We just made it."

"How far apart were the crests?"

Catherine: That's hard to say. Perhaps thirty meters.

"How many times had you capsized since you'd been on the raft?"

Catherine: We'd lost track. There had been three during the first night. Then the most impressive one, in the morning, when our tent had been torn away. Then, during the day, the routine capsizes when the raft was full of water. Were there four or five or six of those? I don't know. We must have capsized eight or nine times, perhaps ten. The last must have been about 6:00 P.M.

From then on, of course, we continued to balance the raft as we had learned to do and we capsized less often. Yet, that was partly because the waves were more predictable and we shipped water less quickly, and partly because luck steered us from the path of breakers.

Lucien: Fortunately. Because, with night approaching

and our fatigue increasing, how much longer could we have held on if conditions hadn't improved a little?

Nevertheless, we had our worst moments that night. We needed sleep, needed it desperately, and at the same time we realized that if we gave in to it, that would be the end. If we fell asleep sitting on the gunwale, sooner or later we would let go of the arch to which we were clinging and fall into the sea. If we stretched out on the raft bottom, in the water, we ran the risk of being completely paralyzed by the cold and also of capsizing, since we would then be unable to balance the raft. The only solution, if we were to survive the second night and see the last of that endless storm, was to overcome the need for sleep and, body-thrust by body-thrust, prevent further capsize. We weren't at all sure we could do it.

I asked Catherine, "Do you think you can stay awake until morning?" She replied, "I don't know, I'll just have to try." Of course, we hoped the wind might die and the sea calm itself during the night, that we might not have to struggle until dawn. But frankly, we never expected any such respite. When I saw the wind wasn't dying at sunset, I thought it would keep up for at least another day.

Catherine: It was agony. The problem was very simple. To survive, one must not sleep. Well, I didn't have the strength anymore to struggle against sleep. During the day I had been able to keep things in hand more or less and to protect myself from the cold, my greatest fear, by repeatedly moving my neck and shoulders. But once darkness came, it was hopeless. I wet my head and splashed water on my face in vain. Nothing worked. The moment we crested a wave, I'd fall asleep, still aware of gripping the arch. I'd wake up when I felt the next wave. We then had to move immediately and put ourselves in the right position to meet

Second Conversation: To the End of the Nightmare

it and, depending on who sat where, quickly lean back or forward.

We did this hour after hour. Our response had become automatic. We no longer had to see the waves or try to understand them to know what moves to make. We did it instinctively.

Whenever I fell asleep, I dreamt that the waves were sinister living things attacking me, trying to snatch me, tear me from the raft, destroy me. I had that nightmare over and over. Perhaps that's why I kept hanging on to the raft even while I slept. The stars were shining, yet the night was dark. The moon was still too new. Except for white foaming crests, the sea was invisible.

Actually, I think the wind began to moderate after 10:00 P.M., making the sea less dangerous if not less rough. Breaking seas were less frequent. It's easy to say that now, but while we were living it, it took us a long time to become aware of it. Nothing had really changed. The danger was the same, and we had to keep up our fight against sleepiness and the cold.

Lucien: We called each other by name, Lucien, Catherine, to keep ourselves awake. We sang "Frère Jacques" together, the only song we could remember.

Catherine: When did we sing and talk? When did I sleep and dream? I don't know. Everything happened at once, as if we were living several levels at the same time. During the early part of the night, it seemed to me I woke up and answered Lucien when he called me, and then continued to sleep while also doing what was necessary to balance the raft. At least, that's how it seemed. Perhaps our contortions didn't do much good anymore. Perhaps it would have been the same had we sat quite still. But as we didn't know, we continued to move back and forth.

I remember Lucien began to talk at one point. It was the first time that he really talked since the second ship had passed us at nightfall—I mean that he said more than an occasional simple sentence. He was talking about death. For him, we had only a few moments left. Something unforeseen was going to happen that would put a definite end to our adventure. His outpouring of despair must have helped me to pull myself together, because I woke up a little and tried to cheer him up. But toward midnight he collapsed. He fell fast asleep clutching the arch. When I tried to wake him, he gave a start and then fell asleep again.

Luckily, from that point on I began to feel better. Oh, I was hardly in top form, yet I'd reached a sort of second state. The wind had definitely moderated—this time I was aware of it—and I was also aware that the sea had a changed aspect. It was choppy and crossed and sickening. I felt we might capsize at any moment. I threw myself right and left, trying to balance the raft. I tried several times to bail with my hands, hoping to remove some of the water we'd shipped. I also pumped up the upper ring and the arch, both of which were becoming flabby again.

Until 5:00 in the morning, I did everything with a kind of unnatural strength, like an automaton. Not only had I completely forgotten about Lucien, to such an extent that I was surprised to see him there at dawn, but I had the impression I was surrounded by people who took a dim view of what I did.

At one point, Lucien woke up. He was furious. He cried, "Shit, when is this going to end!" Then he fell asleep again. It's strange because I remember Lucien's outburst and at the same time I know perfectly well that at that point I had completely forgotten his presence.

At dawn, he let go of the arch. He fell into the sea and

that woke him up. He caught the lifeline by reflex action and managed to climb back alone. I didn't react in any way. I didn't try to help him. To me, he was no longer there.

Lucien: When I'd let go, I'd been dreaming that I was in the raft in a calm basin, a sort of small harbor. On the quay, Pierre, one of my friends, was waving to me. He must have been returning from the market because he was carrying a basket full of provisions. I answered his cries, but the raft stayed motionless in the middle of the harbor.

"And what was Catherine doing?"

Lucien: Catherine wasn't in my dream.

"Still, you managed to overcome the almost insurmountable obstacle of the second night."

Catherine: Yes, we overcame it without realizing it. It wasn't one of those endless ordeals that, at every moment, one thinks one will never last through. I never said to myself, "I can't go on. I'm going to let go. I give up." No, it was a bit like one of those long nightmares that begin as soon as one falls asleep. Even if one wakes for a moment, one is still in the dream and the dream continues as sleep returns.

If we survived, that night, it was doubtless because we still had a desperate desire to live that informed our every move, our every lapse from consciousness, without our making any conscious effort.

Lucien: Yet we weren't saved, by a long shot. My involuntary bath had awakened me, torn me from my dream of the calm harbor. But it left me still somewhat haggard. I looked at Catherine. What could she be doing on her knees in the water in the middle of the raft, bailing with her hands?

It was beginning to be light, but the cold was more

Second Conversation: To the End of the Nightmare

bitter than ever and we were worn out, scarcely able to form a coherent thought. It was a long time before we could pull ourselves together. Until then, we were a bit dazed and didn't even speak. It took a violent movement of the raft to remind us that we were not yet out of this mess, that we were still at the mercy of some mishap, and that it was high time we resumed our watchfulness.

"How was the sea?"

Lucien: The wind had definitely moderated to about force five or six but the seas were still crossed and unpredictable. The raft was almost half full of water, and that greatly increased the risk of capsize. The first thing to be done was to empty it. But how should we do it? Capsizing on purpose was unthinkable. Then we had an idea. I took off my oilskin pants, we tied a knot in each leg and we used them as a container. It worked fairly well as long as we didn't fill it too full. Otherwise it was too heavy and we didn't have the strength left to lift it. At the same time, whenever we felt the raft being swept along by a wave that seemed dangerous, we dropped everything and rushed to the rubber gunwales to balance it. We used the pump to finish the job but, handy as it was for reinflating the raft, it was so hard to bail with that Catherine could hardly work it.

"How often did you have to pump up the raft?"

Lucien: At first, only the upper ring and the arch lost air. We gave the upper valve a dozen strokes of the pump every three hours.

"So, once you had emptied and pumped up the raft, you found yourselves right back where you were before: sitting on the rubber ring, compensating for the action of the waves without the slightest chance of getting some rest."

Catherine: To a certain extent things were as you describe them, yet it wasn't quite the same. To begin with, it was daytime. At night, on the sea, especially in our circumstances, nothing was definite, nothing reliable. Our minds were quite defenseless against the onslaught of hallucinations. With daylight it immediately became easier to fight off the need for sleep which tended to be less pressing then, even when we were desperately tired.

That morning, shortly after sunrise, I really felt awake, as though I'd emerged from a sort of stupor, and had regained part of my strength. At the same time, as Lucien has said, it was becoming obvious that the wind had dropped. It still blew from the northwest and was cold, but little by little we began to feel that the storm was over. The changed appearance of the sea was due solely to the wind. Cross seas, more crossed than they had ever been, testified to the existence, somewhere, of a different wind—from the northeast, no doubt.

Whether it would reach us was the problem. After all, it wasn't out of the question that another tramontane would follow—after a few hours of relative calm—the one that was just beginning to subside. Actually, we were divided between vague hopes and many fears. But from 10:00 A.M. on, the sun really began to warm us up, and strengthened our hopes.

Lucien: By the early afternoon, the change in the weather became noticeable. The wind was down to force three or four. The troughs were no more than a meter to a meter and a half. After what we'd been through, it was almost a dead calm. At any rate, we were no longer in any danger . . . I mean any immediate danger. At last we were able to abandon our position on the rubber ring and sit in the bottom of the raft. Despite the presence of three or four centimeters of water there, this greatly increased our comfort.

Of course, we could have pumped out the water, but it wasn't worth the effort it would have required. We were so drenched that the presence of those few liters didn't really bother us.

Toward 3:00 P.M., the weather turned really pleasant. It was then that we decided to undress, dry our clothes, and soothe the sores that covered our bodies—especially our thighs, arms, and elbows.

Undressing was an extremely painful procedure. Under such conditions, you can't imagine the contortions such an undertaking required. Our arms and legs were stiff and each movement was painful. Once naked, we realized that our bodies were covered with red pimples. Our feet were white, while the flesh under our toenails was red. But what a pleasure not to steep in sopping woolens! We stretched out in the sun and dozed until the lateness of the hour and the growing coolness forced us to move about again. Our clothes were almost dry. Well, not really. But at least they were no longer dripping wet. We had to get dressed again.

Catherine: That was even more painful than undressing and it took us a long time. We did it very carefully, helping each other. We wanted to protect ourselves from the cold. This was still my chief concern; I wasn't suffering from hunger. And, as we drank a few swallows every three hours, we weren't really thirsty.

Lucien: I was beginning to be really hungry. I thought to myself, we won't be able to hold out much longer at this rate.

"What did you feel were your chances of survival?"

Catherine: We talked about that problem, between

naps, while our clothes were drying. First, there was the question of physical resistance. Under the best of circumstances—warm fair weather and allowing for rain so that we might be able to collect water (which were contradictory conditions)—how much longer could we last? I thought ten days, but Lucien was more pessimistic.

Lucien: That's right. I gave us four to five days, no more, but I continued to hope we'd encounter a ship before then that would get us out of our mess. One thing was clear: the ship that had looked for us at the end of the first night hadn't given the alarm. Thirty-six hours had passed. Apart from the second ship the evening before, we had seen no one, either on the sea or in the air. When I remembered that I'd used up my only flare and my three sticks of phosphorous on that damned ship, my blood boiled.

Catherine: I was still convinced we wouldn't be rescued by ship. Where were we? We had tried to estimate our position, but we must have made a big mistake. We had put ourselves much farther to the north than we actually were. At the time, we had thought that if the wind held from the northwest it would bring us to Corsica, where in fact we were making our way toward Sardinia.

While we were aboard the inflatable raft, our estimates varied greatly. But that sort of error wasn't really very important since we had no way of influencing our course. Only one thing mattered. To take advantage of a wind that blew steadily from the same quarter, so that we might be spared an endless shuttling back and forth. The western quarter, we reasoned, meant the sun and the possibility of reaching Corsica. The eastern quarter meant rain without doubt and, if all went well, the coast of Spain. But the coast of Spain runs southeast, and if the wind

veered into the northeast, this would add to the risk of greatly prolonging our journey. In any case, we could do nothing. We had to wait.

Lucien: What should be made clear is that optimism prevailed that Sunday night. An optimism tainted with anguish, to be sure. Yet, we were almost dry, the raft was dry (we had made the effort to empty it), the sea calm, the wind nearly still, and we were going to be able to sleep. We hadn't been so well off for a long time. We had considered taking turns to keep a continuous watch, but it was impossible. We were both too tired. Besides, without flares, we stood no chance of being seen on those dark nights, even if we did meet a ship. And judging by the always empty sea, we didn't seem to be in a much frequented zone.

We lay down next to each other and fell asleep.

II. SURVIVAL

5. Turnabout from the East

They had been asleep since 6:00 P.M. But toward 10:00 P.M., Catherine woke up. She was soaked and frozen, steeping in several centimeters of water that had penetrated her clothing. It was impossible to tell where it had come from during the night. She shook Lucien. "You mustn't sleep, we're lying in water. It's cold." Later, she was to say, "I've never been so cold as I was that night."

Lucien responded to Catherine's appeal by stirring slightly. He was so fast asleep that he had not felt the cold. She persisted. He was furious and snapped at her, "Can't you let me sleep in peace?" But he had to accept the evidence. His clothes were as soaked as they had been during the storm. He sat up in the bottom of the raft, his head groggy with sleep. What was happening? Catherine didn't know. Probably the raft had sprung a leak somewhere. Perhaps the bottom had been torn during one of the many capsizes. In any case, they wouldn't be able to investigate the situation properly until daylight. In the meantime, they had to see to the most urgent matters. First, to bail,

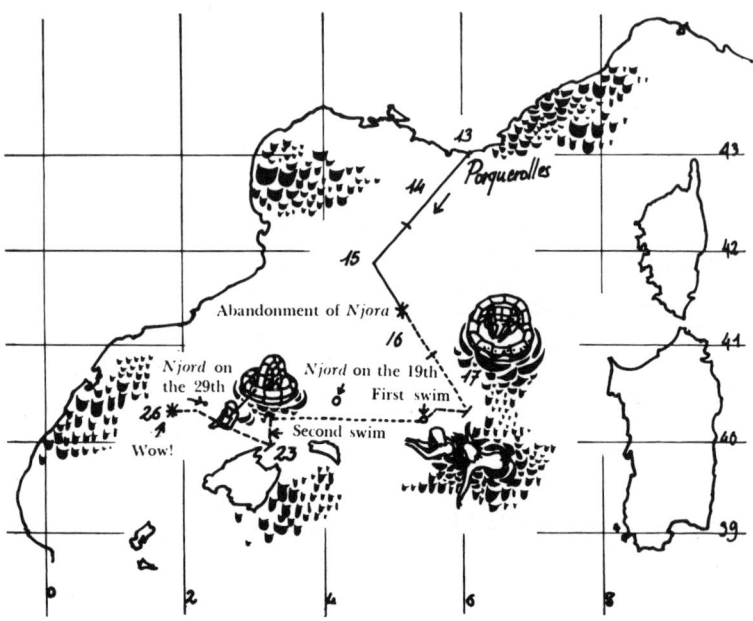

Their course as the survivors reconstructed it after their return.

lest too much water accumulate in the bottom. Then, to find some way to rest as well as possible without soaking in the cold water.

They knew what position to assume to achieve that result. They had already made use of it during the first night, after the first capsize, when they could no longer each sit on a jerry can. They sat with their backs resting against the inflatable rubber rings and their legs crossed under them. This didn't prevent their feet from suffering from the cold or the coldness of their buttocks from reaching their kidneys, but it seemed the best solution. They had already shipped too much water to pump out—it would have taken too long, been too tiring—and there wasn't enough to make bailing with their hands efficient.

Once again it was Lucien's oilskin pants, the legs knotted, that came to their rescue. A few strokes of the

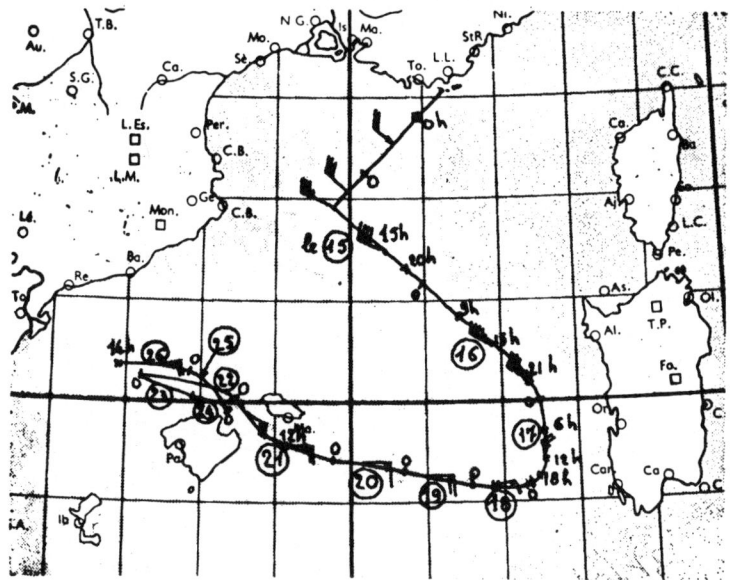

The journey as it was reconstructed by René Mayençon: approximately 200 miles aboard *Njord* and 600 miles aboard the raft, an average of 2.25 knots aboard the raft?

pump finished the job. They took advantage of the occasion to reinflate the upper ring which, after more than four hours, was beginning to go soft.

Four hours, they had slept four hours . . . How long had it been since such a miracle had happened, and when would it happen again? They knew that eventually the sea would break over the raft again; they wouldn't be able to prevent it. So from now on they were condemned never to be dry again, never really able to stretch out again. And to the bondage of periodic pumping up would be added that of continual bailing, with the pump as their only instrument.

It was hardly encouraging to have to look forward to an added complication in their struggle for survival. But they were still too tired for this new worry to have struck home. They would think about it when there was daylight.

For the moment, trying to rest some more was the only thing worth considering.

After trying to combat an insidious thirst with four swallows of water, Lucien was soon asleep again, despite the discomfort of his position. He was dreaming. He was again in that familiar harbor, only this time very near the quay on which his friend Peter was standing. Then he, too, was on the quay. He was walking with Peter. They were going to buy some couscous and a bottle of wine. Now he returned to the raft with the wine and the couscous. He woke up and said to Catherine, "Where did you put the wine and the couscous?"

"What wine, what couscous?" asked Catherine. Still suffering from the cold, she had been unable to fall asleep again. Lucien insisted, "The things Peter just gave us." He stood up. He was still in the boat basin. He shouted, "Peter!" Catherine became annoyed. "There's no Peter here. We're lost in the middle of the Mediterranean." Lucien had to accept the fact. Drunk with fatigue, he staggered, shrugged his shoulders, resumed his uncomfortable position and fell asleep again.

Catherine was dozing. The wind, which had died completely, rose again from the east before the night was over. It began to rain, which woke Lucien. The small of his back was icy cold and the rain was too light to raise any hopes of collecting it. Until dawn, he battled the cold. His oilskin pants reached only to his waist, which left him poorly protected. Catherine was more fortunate; hers rose to cover her chest.

Twice during the night they had pumped out water and had reinflated the upper rubber ring. There was no longer any doubt that the level of the water was rising slowly as the hours went by. With the coming of day, their first task was to make a careful examination of the raft bottom. There they discovered the irreparable damage they had dreaded—a small tear in the fabric. For a moment they were crushed. Of course, they had known it must have been something of that kind, but so long as they were not sure, they had hope of somehow eliminating the cause of the leak.

At the same time, Lucien raged because he had no means by which to move or direct the raft. Now that the wind hardly blew, it merely bobbed in place. Had the paddles not been lost during the first capsize, it would have been possible, thanks to a ball of string discovered in one of the raft's pockets, to use them to rig a mast. They could have improvised a sail from the remnants of the tent . . .

While mulling over that regret, the idea came to Lucien that he could try putting together the pieces of the fabric and lashing them to the arch. Catherine and he worked a long time to fashion a kind of laughable spinnaker, thanks to which they had the impression of making some way.

To take full advantage of their jury sail, they had to do away with the antidrift pockets that dragged against the movement of the raft. It seemed a difficult operation. Even if they were able to remove the pockets, how would they right the raft if they were to capsize again? Until now, they had depended on the pockets for this purpose.

From then on, their course was due west. They had the impression that it was a fatal direction for them, that they were doubling back over waters already crossed.

"A west wind, an east wind, we'll never get out of this," said Lucien.

They had no hope of raising the Spanish coast. It was, they reasoned, much too far away. Yet at the same time,

since nothing seemed certain, since the rage to survive was foremost, they refused to make any final analysis of the situation. "Perhaps we're wrong. Perhaps we're closer to land than we think." And then, there was always the chance of rescue by ship. A ship, a landfall! Their will to live was pinned to those two hopes, Catherine favoring the first, Lucien the second.

But the sea was empty, starkly empty, gray and animated by a slight ripple. During the morning it had rained a second time, harder than during the night but not enough to allow replenishing the water supply. There were approximately four liters left. Yet, for the time being, awareness of the extreme short supply of their last resource didn't pierce the envelope of torpor that insulated them from their agony. To live meant primarily to doze, to take full advantage of the chance at last afforded them to abandon themselves to the moment without always having to struggle to the point of exhaustion.

Half stretched out, half squatting, they continued to catch their breath like swimmers who remain lying on the beach, having feared they could not overcome the current that had been dragging them out to sea. And the anaesthesia that cut them off from reality was reassuring. Something about it promised a future, a strange, blurry future, like the world filmed through a drop of water, yet one which existed just the same.

From time to time they emerged from that cotton-wool world from which even dreams seemed to have fled. They spoke of *Njord*. What could have happened to her? Lucien said that, left to herself, she must have been tossed about by breakers and have filled little by little until she had gone down. They hoped that was what had happened. If the boat were discovered with no one on board, they didn't want to be reported missing before being found.

They thought again of those hours spent battling the sea, running before the wind, the ripping of the storm jib, the sea anchor, the first capsize, the huge wave that had submerged them. And the sequel, the panic that suddenly had seized them and led them inexorably to try inflating the rubber raft and, having done so, to take refuge aboard it.

After that, everything was blurred—a mass of recollections from which nothing emerged but a feeling of unreality. Everything they had lived through during those thirty-six hours while the storm had been upon them seemed like an absurd episode, almost outside time, unimaginable: a demented dream from which to have escaped alive was incredible.

Whenever they began to evoke the period between the beginning of the tramontane and the moment when they had lain naked in the sun along the bottom of the raft on a sea calm at last, their recollections tended to focus on one last image—that of *Njord* disappearing into the night. Everything that had followed had scarcely touched them. Nothing remained but tremendous lassitude and a passionate desire to escape from their liquid prison.

Their feet were white and icy cold. "The feet of corpses," Catherine was to say. From time to time one or the other felt a fiery pain in the heel. It came all at once, was unbearable, then little by little got better. They held their feet in their hands to warm them up, then hit upon swaddling them in the raft's sea anchor, a sort of parachute about forty centimeters in diameter.

Night overtook them this way, sitting side by side, on the whole fairly rested, with their feet in the sea anchor. It was their fourth night adrift, their sixth since leaving

Porquerolles. Catherine engaged in lengthy calculations to make sure that she had not, somewhere, overlooked a period of time. No. After having reckoned and reckoned again, she was sure of her facts from then on. It was indeed the evening of Monday, the 18th of September, and they were drifting slowly west. The night passed without incident, marked as usual by sessions at the pump.

The morning of Tuesday the 19th was clear. The wind had backed into the northeast. Airplanes passed above them, far too high, of course, to give their occupants the slightest chance of spotting them. The airplanes were heading north. It was a curious feeling to imagine men and women seated comfortably in those cabins, surrounded by hostesses ready to fetch them whatever food or drink they might want.

Every now and then the passengers would look distractedly through their windows at that expanse of water over which they were flying. In a few moments they would be landing in Marseilles, Nice, Paris. They would hear the noises of an industrialized world—the shriek of jet engines, the falsely confidential voices coming over the airport loudspeakers, car motors, the buzzing of the crowd. They would remain in a world that rules out chance. One left at such and such a time, one arrived at such and such a time, a meal awaited one at a given place, one's work at another. One went toward one's home, one's family, one's habits. One risked nothing, except the unforeseen. Yet everything, or almost everything, had been foreseen.

They, however, were in that rubber raft, forgotten, totally at the mercy of chance, that chance so carefully eliminated elsewhere. The distance separating them from some shore would seem absurd to the occupants of those great airplanes: barely a few minutes' flying time, the time to drink some fruit juice, to nibble a sandwich, to exchange a private smile. But there, on the sea, even allowing a continuing east wind, it would take Catherine and Lucien days to reach a haven. And during that time, airplanes might fly over them, ships pass them. No one would care about them, no one would look outside his own world. Everything was happening as if they had been propelled into a fourth dimension from which they watched everything without anyone suspecting their existence.

Urinating posed serious problems. They were astonished that, drinking so little, they should continue to pass water with such regularity and to void in such large quantities. Yet, each time, they had to confront the painful and exhausting tasks of undressing and dressing. In the beginning, during the storm, they had no other alternative but to wet themselves. Soaked as they were, it did not strike them as important. But once pimples had begun to appear on their skin, they had had to abandon this easy practice: it had resulted in such burns that the painful procedure of undressing finally proved preferable. If, because of excessive fatigue, they failed to perform this task, the punishment was terrible. Each time, it took Catherine half an hour to undress and left her exhausted. Yet it was also reassuring, for a kidney stoppage was one of the gravest dangers threatening them. It was the reason they dared not drink even small quantities of sea water.

The weather improved. Tuesday the 19th was the first time since the 14th, the day the tramontane had started, that the weather had been really warm. "Let's bathe," suggested Catherine. "We could dry our clothes. It would be pleasant."

Why not? The wind was so light that the raft scarcely drifted. The water around them had the opaque translucence of great deeps sleeping in the sun. Of course, the problem of undressing arose again, but after that exertion there would be the reward of being naked in the clarity of warm sunlight, almost still air, free of that heap of brine-soaked woolens.

Slowly, cautiously, painfully they freed themselves from the grip of pants, pullovers, shirts. They slipped into the sea, swam several strokes, lay flat on their backs, and stretched. Their muscles flexed again as if they had just been released from some confining matrix. The feeling of relaxation was extraordinary. "For a moment, I thought I was on holiday," Catherine said later.

Fish were swimming beneath the raft. They tried naively to catch one. Using the knife on board the raft, Lucien busied himself cutting off the antidrift pockets. The main thing was to go as fast as possible, wasn't it? It was a hard job but he managed it.

They climbed back aboard. To get some of the salt out of their clothes before drying them, Catherine rinsed them in sea water. Then they basked in the sun. They were no longer two shipwrecked souls clinging to a vague dream of rescue, but a man and a woman who watched the spots the sun made under their closed eyelids.

Reality, however, was not altered by that moment of well-being. It would soon reassert itself in the form of sticky, half-wet clothing, still saturated with salt despite its peculiar rinsing, that would have to be donned over sore skin and limbs grown thin. Not counting the few mouthfuls of corned beef they had eaten, they had been virtually without food for five days now.

It was night again, the fifth since they had been in the raft, the third to have begun on a calm sea. Calm? That remained to be seen. After 9:00 P.M. the question began to

Turnabout from the East

arise. A big storm was developing to the north, while the wind, due east, was rapidly rising to force four or five and seas were beginning to develop. But to them, the storm did not signify the danger of a gale; it signified rain and the possibility of collecting water.

They were seated side by side facing their "sail," thanks to which the raft no longer spun in circles. It now had a bow and a stern to offer the wind. The storm moved around them and lightning lit up the sea. The sight was impressive but they had no thought of admiring it. They hoped for rain, for water. Their only fear was that the storm might leave them without having poured its torrents upon them. For almost three hours they kept track of where the lightning came from so they could follow the course of that mass of stormy clouds. Now and then they felt a few fat drops fall without being followed, however, by even a drizzle.

Then abruptly, after an endless wait, there was an avalanche of water, a deluvian downpour that came to rinse their faces, their hair, their hands, streamed down their oilskins, filled the bottom of the raft. Lucien tried drinking the water. He spat it out. It had a dreadful taste. Before hoping to collect water for drinking, the interior of the boat would have to be washed and rinsed. But first they would have to drink, drink their fill. Lucien rigged a strip of the tent in such a way that it caught the rain. They drank for a long time. The water was very pure, very cool. A true nectar. Then, using the pump, they emptied all the water that had accumulated in the bottom. It filled again rapidly.

The rain never seemed to stop. It fell for nearly three hours. They were able to collect only a dozen liters, somewhat briny. But with these, together with the four liters remaining in the jerry can, they had provision enough to last them several days.

They were happy. Their happiness had almost made them forget their fatigue, although throughout that entire period of activity, they had continued to weigh their every move, rather like a cardiac patient obliged to measure every effort. Besides, it was exhausting to do anything on board that raft; if they were neither seated nor lying down, every move raised the problem of balance. For example, while they were collecting water with the pump, they nearly upset the jerry can several times.

When the rain stopped, they began to feel cold. The coolness of the water which, a moment ago, they had found so pleasant as they drank, now penetrated and chilled them. "The rainwater must have been about ten degrees centigrade," * Catherine estimated, "whereas the sea was about eighteen degrees."

They tried nestling against each other to warm up but the encumbrance of sweaters and oilskins prevented that method from being effective. The storm rumbled in the near distance while the wind continued to blow at force five with every now and then brief, more violent gusts. They were lashed by spray. They might have to endure another downpour or one of those squalls that often empty thick, heavy clouds. One thing reassured them. In such a case seas would scarcely have time to build, and wind without waves was a precious ally. While pushing the raft toward land, it would also take it from that deserted zone which no ship seemed to frequent. Perhaps it would carry them as far as the lanes of freighters and liners.

Now that they had drunk their fill and replenished

---

* 50° Fahrenheit.—Ed.

their water supply, they wanted above all to be able to rest, to warm themselves up and sleep a bit. More rain would mean more cold and having to pump again. It would be bad enough to have to perform this drudgery two or three hours later, as was usual. Fortunately, the storm went away and their clothes slowly reached the temperature of their bodies again. And the night was calm, as they had hoped.

6. Third Conversation: The Wait

Lucien: Wednesday morning was the "anniversary" of our departure from Porquerolles, just a week before. The weather was good, at least fairly good—sunny in spite of a layer of fog high in the sky. A fresh wind still blew from the east. We had the impression we were making headway, maybe a knot, a knot and a half.

Planes kept passing overhead, always heading north. We had been under that air lane for more than twenty-four hours. So I began to wonder whether that "impression of making headway" was an illusion, whether in reality we had remained for days and days practically stationary in the water. How could we tell? The only reference point that we had were the planes themselves. Yet it was true, as Catherine said, that we had moved, for there had been no planes before, just as it was true that our progress was so slow that it was natural to see planes now for a long while. My doubts were absurd. Nevertheless, they preoccupied me for a long time. It shows how subjective, uncertain, confusing everything becomes when one is lost at sea, as we were, without any bearings to hang on to.

Third Conversation: The Wait

It's different on board a boat. First, there's the effort the hull makes to drive back the water, to push its way through it. Forward movement is soon apparent. And then there are instruments of navigation. The miles inscribe themselves magically on the dial of the log. One marks one's course on the chart. One can see there in concrete form the point one has reached and the point one is steering for. A mysterious thread stretches between the chosen point of arrival and the compass card that joins the boat to its destination. Finally, when in doubt, there is the sextant which lets one consult the stars about the subtle changes which the distance run has worked on angles the naked eye cannot perceive.

None of these were available to us. Due to the raft's negligible displacement, our drift was scarcely noticeable. There were no eddies. I found myself watching the water's surface for long periods, trying to find a floating object of some sort, a piece of seaweed, a bit of wreckage, some flotsam, a bubble, whose movement in relation to us would confirm that we were, in fact, making way.

I regretted not having things to throw into the sea, pieces of paper, for example, just for the pleasure of watching them drift past us. There was another solution, that of spitting into the sea, but we didn't have enough water to waste our saliva in such a way. Besides, spitting was only really effective in very calm weather when our speed was near zero.

Another worry: would the wind stay in the east? Sunday evening, at the time of the capsize, we had been very discouraged for a while. Our hopes of reaching Corsica were crumbling. That hope had been quite fanciful, but I didn't realize that until much later. We had thought it was possible, before the wind had risen in the east. But after three days of drifting westward, the prospect of the wind's shifting again and sending us back to where we had come from, to where we knew no ships passed, terrified us.

Our fears increased as we came to realize that only the east wind was likely to bring rain and hence drinking water. Besides, we could not face the possibility of having to bear up under another tramontane. Three days of east wind meant the end of a cycle now. What would happen the following day?

"What would you say your position was then, let's say at the end of Wednesday, the twentieth?"

Catherine: We were at the same latitude as Barcelona. It was the one thing we were almost sure about. And yet our sureness was rather shaky for, aside from the fact that we had no idea of our longitude, our overall estimate varied considerably depending on the moment and, especially, the level of our morale. At the end of the storm, as we've said, we thought we were a hundred miles or so from Corsica. But when we say "we thought" we really ought to say instead "we hoped." Our reckoning was far more subjective than scientific.

In fact, I suspect that without wanting to admit it to ourselves, we adopted the solution that seemed to us most likely to offer a chance of rescue. If the wind blew from the west, we were near Corsica. If it blew from the east, we were as far west as possible. Thus, as time went by, we accumulated more conflicting estimates and were less able to form an exact idea of where we were, which gave us considerable margin for constructing hypotheses that corresponded to our moods. Here's proof. We actually wondered at one point whether we were to the north or to the south of the Balearics . . . While all the while we

Third Conversation: The Wait

held to our belief that, in all possibility, we were bearing down on the Costa Brava.

I had often tried to reconstruct our journey by drawing with my finger on the inflated gunwale of the rubber raft. I attempted to locate the coast of France, Corsica, the coast of Spain, and the Balearics. We were no doubt somewhere inside that quadrangle. But where? I tried to make an estimate: two days aboard *Njord* heading southwest, then running southeast for eighteen hours. How far did we go in our raft, and in what direction? Due east or was it rather southeast? And how far had we gone since we had been heading west?

At one point, despite my efforts, all the data got mixed up. The results I got were completely contradictory. Moreover, what did I have to check them against? What relationship to reality could my imaginary map have had? Lucien and I argued about it so much that Corsica, the Balearics, and the coast of Spain began to occupy mad positions.

Several times we even failed to agree on the meaning of the cardinal points of the compass. East took the place of west. The sun made bizarre journeys between unexpected risings and settings. It was as though the earth had begun to revolve in the opposite direction just to baffle us. Having reached that point, there was nothing left for us to do but to take refuge each in his corner to dream and reflect in peace. We no longer had the strength to pursue fruitless discussions.

Lucien: Sometimes we made plans for the future. As for me, I had never stopped making them. From the beginning, during the storm, I'd spoken of buying a minibus to see Africa with. I think I've mentioned that. I'd thought about that minibus ever since then. Occasionally

I would describe it to Catherine, explaining how we would go about fixing it up, what we could do with it. But I felt that she wasn't very enthusiastic. She agreed only reluctantly to discuss the possibility, until the day she began to tell me about that farm that she had been dreaming about.

At first, I was rather against the notion of a farm. The idea of seeing the same landscape every morning when I opened my window, the same tree, the same field, the same mountain, was opposed to everything I had wanted to do until then. Yet, little by little, I got caught up in her fantasy. I began to live on this farm. We made our own bread, our own butter there. Yet every now and then the desire to leave overtook me. I found my minibus again and was on the road once more . . .

"Why this constant need to escape?"

Lucien: For a year I worked in a shoe store. It was hell. People would come in. A hundred pairs of clodhoppers stared them in the face. They always wanted the hundred and first pair. I couldn't stand it anymore. I had to leave, to travel, to try to find another way of life while I was still young enough to do it.

Catherine: As I said before, I'd traveled a lot, with my family and alone, since the age of fifteen. The exotic no longer really attracted me. For example, when people mentioned Tahiti or the Pacific to me, it made me smile. What more could I find there than elsewhere? I knew the only thing that counted when one traveled was to meet new people.

Every time I'd been somewhere and hadn't known anyone, I'd had no interest in the place. I'd sometimes dreamt of long journeys on horseback, giving me time to mingle with as many people as possible. One of the things that I'd liked about sailing was the world, often strange,

Third Conversation: The Wait

in which one was immersed; the navigators, real or would-be, those who set sail and those always about to set sail, those who arrived and those one didn't hear of anymore, the strange fauna of sea vagabonds.

"Yet, the dream of the farm was more the dream of an isolated life."

Catherine: Yes, but at that moment, in the raft, I felt above all a need for security. After all, it must be understood that at no point did our projects, however precise they may have been, succeed in making us forget our plight. We never said: we're going to do this or that, but always: if we are saved, we could consider doing such and such a thing. That's very important. It must be rare to be in a position in life where one can really never say: I'm going to do this or that. I think even a prisoner has a greater margin for initiative than we had. At least he can try to escape, wrestle a guard, climb a wall, even if it means being recaptured.

By contrast the range of our decisions was extremely limited—whether or not to drink, how to arrange ourselves to try to minimize our discomfort, and, finally, suicide. In short, we were quite powerless in the face of the future, whether near or distant. There was only one thing left we could do if we wanted to regain our freedom one day. We must try to survive. And this fact struck us— struck me, at least—with particular force whenever we summoned up some project. I quickly rebelled against the absurdity of any such dream. We had no choice, we must hang on, survive, and therefore husband our strength, save all our energy for the struggle against exhaustion, pain, the cold. Above all the cold for, except for a few rare moments, we had never been really warm.

We thought out every move ahead of time, held it back, didn't make it unless we felt it was indispensable.

We ended by adopting the same attitude toward speech. We had formed the habit of exchanging only a few words every now and then. Not true dialogues. An idea, an image, a memory that one of us would utter and let hang in the silence, and each made what he wanted to of it.

Lucien: Yes, and the idea put before us that way on that day was the idea of death, my death, my slow death. Oh, I'd already thought about death, in the beginning, at the time of our struggles with that sea gone mad. Then it had been a question of violent death. We were being tossed, battered, overwhelmed. This time it was no longer the same thing. Each saw his own death written in the face, traced in the silhouette of the other—the features deepening, the eyes sinking, the body shrinking. Under our oilskins, we saw each other grown thin, insubstantial. Each of us saw in the other our own disappearance. Yet,

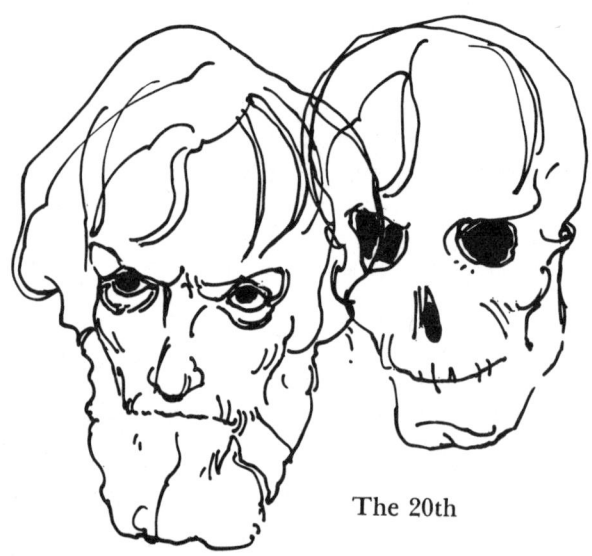

The 20th

at the same time, trying to persuade oneself that the other was wasting away more rapidly. One wondered: "How am I? Am I not in better shape?" And sometimes one of us would ask the other, "How do I look?"

Catherine: Not having a mirror bothered me greatly. I tried to see myself in the back of my watch or in the knife blade without much luck.

"Did you still have dreams?"

Catherine: Yes. In fact, the moment we dozed we were elsewhere and we sometimes had trouble, on awakening, returning to reality. I remember two dreams that I must have had about the time of the 20th of September. The first was about one of our friends who had come to see us in a small motorboat and who left to get help.

Why didn't he take us on board his boat? I admit the question never crossed my mind. In my confusion I must have thought it was necessary to undergo a number of transitions, trials, and rites before leaving that state of living death and reentering the incredible world in which it was possible to meet other people and to decide what one was going to do immediately afterward. Rather like what the ancients had to go through in order to return to the earth from the kingdom of the underworld. The friend's visit represented to me the first step in the direction of freedom, a kind of encouragement. The rest would follow.

In the second dream, Peter, that same celebrated Peter with whom Lucien had thought he had gone to buy some couscous, came to dine with us aboard the raft. When he went, he left a bottle of wine, white wine. When I woke up, it was my turn to ask, "What did you do with the white wine Peter gave us?"

Meanwhile, the sea was still deserted. In more than

five days since we had abandoned *Njord* we had encountered only two ships, one at the end of the first night, the other twelve hours later, on the evening of Saturday the 16th. Since then, nothing. We saw the third ship at dawn on Thursday the 21st. But it passed far away, more than a mile away. The weather was foggy and gray. The wind was rising. A slight drizzle was falling. It was impossible to signal for help.

Later, during that Thursday, the wind began to blow harder and harder. It must have risen to force seven. The seas were well shaped, long and regular. Our embryonic sail was working, and we raced from one wave top to the next. We must have been making at least three knots then.

"Weren't you afraid of capsizing again?"

Catherine: No, not really, at least during the first few hours. It wasn't at all the same kind of sea as during the storm. We didn't have the impression of being in danger. On the contrary, we were happy. We were aware of going fast, of approaching our objective. After the torpor of the preceding days, we were seized with a sort of excitement that seemed to infuse us with new strength, so much so that at one point that day we thought we had seen land.

Lucien: Yes, on the horizon, toward the west, I seemed to make out a light brown strip. We stared hard for several hours trying to make out something. We already saw ourselves on a beach, if only the wind held.

Catherine: If you stare long enough at the horizon, you begin to see anything. After a while we had to accept it was an illusion.

"How did you use up the few liters of water you had?"

Catherine: We drank an average of seven to eight mouthfuls every two or three hours. Yet it wasn't actually

Third Conversation: The Wait

rationing. We hadn't imposed fixed rules on ourselves. Every now and then one of us would suggest we drink. Then we would either drink or decide to wait a little. It depended on how we felt.

"And what about hunger?"

Lucien: For me, at least, it was a permanent obsession. While wide awake, I would see my mother preparing a huge jar of chocolate mousse or pouring me a glass of milk. But, since she didn't see me extend my hand to take them, she would put them away in the refrigerator.

Catherine: My God, did I have to listen to him talk about his mother's chocolate mousse! It annoyed me even more because I was trying not to think about the problem of food.

Lucien: That day, I continued to stare at the horizon without saying anything until evening. The business about land continued to go through my head. I wasn't really convinced I'd been mistaken. Toward the middle of the afternoon, we began to see gulls, which might mean that land was not far away.

Catherine: The wind continued to blow around force seven. The sea was gray-green, full of whitecaps, hardly appealing. Spray had begun to fly again. It struck us in the face while it filled the raft's bottom, obliging us to pump more often. It was not unusual to ship a sea at the very moment we had finished pumping. All of this drained our strength and sapped our morale. We were very worried because the pump was becoming more and more difficult to operate.

"And what about *Njord*?"

Lucien: I had reached the point at which I thought more about the food remaining on board than about the boat itself.

"Had you thought that you might find her again,

might see her suddenly floating not far from you? It could well have happened for, as we later learned, the *Njord* was barely a few miles north of your position on that day, Thursday the twenty-first of September. You could very well have seen her again."

Lucien: Yes, I did think something like that could happen . . .

"But had you also thought that, even if you were to see your boat again, nothing short of a miracle would let you overtake her?"

Lucien: No. I think that had I seen her, I would have stopped at nothing until I had managed to rejoin her. I would have swum after her if necessary. Although on that day, of all days, it would scarcely have been easy. The wind was freshening more and more and the sea was again becoming rather dangerous, so much so that we began to wonder whether we weren't in for another series of capsizes. We were shipping water rapidly, and had to pump more often, obliging us to make further painful efforts.

Catherine: Also, we no longer had the strength to stay seated on the rubber ring and balance the raft as we had done in the beginning. We squatted in the bottom. We suffered increasingly from the burns all over our bodies. In places, we had wounds that were beginning to suppurate. Lucien had an open wound on one of his heels. I could see red flesh but it wasn't bleeding. Moreover, he hadn't realized he was cut there. I was the one who pointed it out to him.

Lucien: In the end, it was the rain that made the wind drop that evening, a downpour, worse than the thunderstorm two days earlier. We were able to fill the jerry can that time. Because we found the contents a bit too salty, we even allowed ourselves the luxury of half emptying it, and then filling it again. And then we drank with even greater frenzy than the time before. At least, I did. I took so much water that I felt drunk, I was staggering. I must have sat down in the bottom of the raft to catch my breath. After a moment, I began to drink again. I could not believe it was possible to lose so much water.

Catherine: Yes, but after a while we began to wish it would stop. The first rainstorm stopped shortly after we no longer needed it. But now the freezing rain kept on pouring down. It was more than we could bear. And at the same time, it drove us mad to think we couldn't save more of it. Oh! if only we'd still had the second jerry can. The raft was filling up. We were drenched, chilled, hunched over whatever warmth was left in the marrow of our bones, offering our bent backs to the downpour.

That time we were rinsed. There was no question of any salt being left in our hair, on our faces, hands, or oilskins. We couldn't have done a better job at home in the garden in the sun with a watering hose. But the salt that had accumulated in our sweaters and shirts and pants, that burned our skin, dug into our wounds, made cracks where our limbs bent, was not touched by those cataracts of sweet water. It stayed there, under the oilskins, sticking to us like glue. We would have had to undress, to really shower, to rinse out everything, but it was impossible to do so at night. It was too cold. We never would have lasted till morning had we done so. There was nothing else to do but stay still under the rain and steep in our briny clothes.

"And when the rain stopped?"

Catherine: We continued to wait. We'd be at a loss to say exactly what happened that night. We had reached the point where time, space, our sensations, the day before, sleep, reality, and dreams were nothing more to us than a

sort of grayish soup in which we were immersed and from which it was impossible to emerge short of some major event.

For the two days that followed after the rain stopped Thursday evening, everything remained very muddled in our minds. We could never say for certain at what point something happened and whether it took place before or after something else. We began to enter a sort of second state from which we emerged now and then only to sink back again immediately. I've already described it—a sort of grayish soup. We will try to recount what happened from the end of Thursday the 21st until the morning of Sunday the 24th, but our story runs the risk of being blurred, confused, even incoherent.

7. When the End Has Come

Was it indeed during Thursday night, a few hours after the rain? For a long time Lucien and Catherine were divided on that point and then finally agreed. It was indeed on that night. They were sleeping. The noise of an engine awoke them. They sat up. Behind them in the southwest, less than 400 meters away, was the enormous dark bulk, crowned with light, of a liner making straight for them.

No way to escape that threat. The wind had abated. The rubber raft bobbed gently on the remaining swell, the only reminder of the preceding day's heavy easterly blow. They had the impression of being nailed to the very spot about to be leveled by a steamroller.

The ship was bearing down full speed. If her course were not altered one way or another, if they stayed a few seconds longer in the path of that bow, they would have no other recourse but to jump into the sea to try to escape being run down.

They held their breath. They were already overwhelmed by the noise of the engines, then the roar of the bow wave, while the wall of steel, having missed them by several meters and having failed to capsize them, began to slide past them. They cried out, shouted, waved their arms in the night. The ship was long, one of those large ferries that carry holiday makers and their cars to isles of happiness. The sweetish smell of hot oil invaded the raft. And then came the stern and the wash of the propellers. It was over. They had escaped the danger but had lost a further chance of rescue.

Exhausted, they fell back into the bottom of the raft. "I even saw a sailor," Lucien said later, "walk along a gangway, open a hatch, and disappear inside the ship." Catherine did not confirm having seen anything of the kind. Perhaps Lucien was confused. Perhaps it had been several hours later, after daybreak, that he had seen a sailor aboard another ship that had passed close by—less close, certainly, than the one during the night, but nevertheless within hailing distance, it had seemed to them. Close enough in any case to have triggered anger instead of despair when they realized that no one had seen them. They raised their fists against the crew in a gesture of rage.

Earlier, just before their great fright and no less great disappointment, they had seen two other ships pass about a mile away during the night. They were all going in the same direction, northwest. Perhaps there had been others. They would never know. They no longer had the strength to keep watch.

Catherine was dreaming. She was in Paris, at the home of friends who had asked her to dinner. She was telling about her adventure. She was explaining how she had been saved after spending twelve days on the rubber raft. How had she been saved? She did not know, just saved. And Lucien? She did not know what had happened to him. Lucien was not part of her dream.

Saved? By a ship? She believed in that less and less. "A ship would have to run us down, in broad daylight, in good weather . . . Of course, it would be a different story at night, if we had flares. We might stand a chance of being seen then . . ." There remained the chance of reaching land, of being washed up on some beach, of walking on the sand, undressing, drying oneself, catching one's breath . . .

But land could also be a sheltered cove, a rocky point, a sheer cliff, with breakers dashing one against rocks. It could be a friendly village or a deserted coast with the nearest house, the nearest help, the nearest drinking water kilometers away and without so much as a path to show the way. Catherine knew all that, of course, but she drove it from her mind, just as she would not admit another fact: they were both so exhausted that they would scarcely be able to take more than one or two steps on dry land. For her, land had to mean only the assurance of life and not the beginning of salvation.

She refused to despair. Something deep inside her, and beyond all evidence of their predicament, knew that the nightmare would end well. Without quite daring to admit it to herself, she felt she was invulnerable. "I'm one of the lucky ones. And I'll prove it again this time. I know I'll come out of this."

She had dreamt that she was speaking of the twelfth day. The dream had greatly impressed her. How many times had she dreamt that dream? She no longer knew. It was a part of her. When would the twelfth day be? She began to wrestle again with her slow and prudent reckoning of time. The first day had been Friday the 15th, exactly a week ago. It was Friday again, wasn't it? Yes, there was no doubt of it, it was Friday again, the 22nd, thus the eighth day. That left four more days to endure, in fact five if one counted today, which was far from over. Five days and how many nights? Four, the twelfth day would be Tuesday. It was a long way off, a very long way. Would they have enough water to last until then? Five days at three liters a day, one and a half liters per person, that's just about the ration, that makes fifteen liters. The jerry can must hold eighteen. No need to worry on that account so long as they didn't let themselves go, didn't give in to the constant temptation to drink their fill.

Catherine stirred a bit, trying to find a more comfortable position. She had the feeling the burns covering most of her body had grown worse. She was unable either to sit for long, because her buttocks were so sore, or to lie down, because of the water. It was impossible to kneel for long because one lost one's balance with each movement of the raft. Thus, she was condemned to be uncomfortable, to be in pain, to gnaw at her misery in her corner, without hope of solace. Her hurts would only grow worse with time, as the burns had and also the sudden stabs of pain in her heels which brought forth a cry each time, and which occurred ever more frequently, for Lucien as well as her.

Poor Lucien! His morale was certainly low that Friday, almost as low as his physical condition. She watched him dozing opposite her. Under his thick beard his face seemed reduced to a skeleton. His eyes had sunk within their sockets. His oilskin flapped about his emaciated frame, his hollow shoulders. A moment ago, he had roused himself briefly from his torpor to speak of death, of a death he felt was unavoidable if not imminent. "This time we're done for, we'll never make it. It's impossible. We're too weak. It would take a miracle."

Yet it was precisely in a miracle that Catherine had put her faith. She had tried to persuade Lucien that it was still possible, but without much success. She was, however, hardly worried about that. First of all, in their present state, the other person no longer mattered much. It was each for himself, in his own corner, that made it possible to continue the fight to survive. Secondly, she knew that a sudden rekindling of hope and optimism could follow acute despair.

Such an attitude was typical of Lucien's behavior since they had lost *Njord*: a zigzag profile made up of peaks of optimism followed by valleys of profound despair. No sooner did something happen on which they might pin their hopes—the sun bringing warmth, the rain bringing drink, a ship appearing on the horizon—than energy was found again, confidence reappeared, life once again hummed with its procession of dreams and projects. The disappointment that followed such a peak was acute.

Catherine had rejected that continual seesaw between existence and death, those excursions into a future that left one crippled once the future failed to appear and was replaced by a wall against which they butted their heads, a wall that separated them from their lives, prevented them from building a tomorrow. Catherine, falling back upon her own resources, steeled herself to wait. She had made herself stingy with movements, dreams, hopes. To hang on, to hang on, to cling to one's last ounce of strength. Then, when the right moment came . . .

Lucien was stirring in his sleep. He, too, constantly sought that nonexistent position of comfort that would not quicken his burns or rouse the thousand aches that racked his body.

He opened his eyes, looked at Catherine, asked her, "Have you had your shower?" She looked at him with some alarm. "What shower? What are you talking about?" He insisted. "You know, that shower, before the meal. I took mine. They told me they would give you one."

She shrugged her shoulders, preferring not to reply, but Lucien persisted. "Well, look around you, Catherine. You know perfectly well what happened. That big ship came. They opened the after hatch. They brought in the raft. Then they took me to the shower and they said they were going to serve us food. Now, you must go have your shower. It will do you good."

She became angry. She was worried and annoyed by Lucien's behavior. He seemed to see more than the unchanging universe they were trapped in the midst of, the gray and empty sea, the inflatable arch and rings of the rubber raft that were slowly leaking air (and would soon need pumping up). She replied brutally. "Stop behaving like a fool. You know damn well there's no ship, no shower, no meal. Nothing. No one. We're alone, as usual."

Lucien was not affected by her withering outburst. It was Catherine who was mistaken, who did not see, who failed to accept the evidence. They were in the womb of a great ship. Those who had saved them would soon reappear. They had gone to fetch food. "You ought to take a shower," he advised gently.

She was exasperated. She wanted to shake this man whose mind was wandering, who, wide awake, was living some dream. She almost shouted, "There's nothing, I tell you. You're delirious. Pull yourself together. Get up. Have a look."

Something in him responded to the voice that castigated him. He was about to reply but the words stuck in his half-opened mouth. Slowly, painfully, he rose to his knees and contemplated the sea around him. Had the ship already left?

"No ship," repeated Catherine, "no rescuers, no shower, no meal, nothing."

"While Catherine was speaking to me," Lucien later explained, "I was aware of being in the raft, in my usual place, but I didn't want to admit it. I was convinced that the other impression was the true one, the one in my dream. It was as if I'd had a choice between two realities and had opted, naturally, for the better one. It took me a long time to accept the truth. And once I had admitted that Catherine was right, that the sea was empty, I was completely crushed."

It was raining again, a sort of icy drizzle, too fine to drink but wet enough to soak everything. It rained off and on like that throughout Friday. It may have been that additional trial, the cold that little by little enveloped them, that provoked the kidney pains from which Catherine suffered the following night.

Never until then had she felt so miserable, so weak, so vulnerable. Sitting had become unbearable, yet there was no other possible position. For hours at a time she tried to escape from that new pain while Lucien, although numbed by exhaustion, spoke in his sleep and tossed about from left to right, at the whim of the swell, the length of the half-deflated raft.

She said, "I'm in pain." But there was no one to hear her. "Is this the end? Is each of us going to retreat into the ultimate confrontation with death? After all, why not?

Death needn't be hostile. It's a way out. No more pain, no more struggle. To fall asleep at last without having desires harass one in a dream; to set aside the burden, to surrender . . ."

The moon rose. It was almost full. Its presence was reassuring, if only because it gave shape again to their surroundings. A vibration in the water roused Catherine's attention. She could feel it through the bottom of the raft. Instantly, the kidney pains left the center of her thoughts. Something was happening. Then she saw, straight ahead, the running lights of a boat that seemed to be heading their way. The lights that made it visible were relatively low in relation to the surface of the sea. No doubt a motorboat. Now she could actually hear its engines, not just feel them through the water. She woke Lucien. "A boat." With the moon, it was possible that they could be seen.

The news had roused Lucien from his apathy. He glanced around him and suddenly there were three boats in less than a square mile, and one of them, the one which had just attracted Catherine's attention, was indeed heading toward them. It was a motorboat, a large motorboat, thirty to forty meters long, and it was going to pass less than a hundred meters away. The police, the customs, some millionaire's motor yacht? There was no light on board except for the running lights. A phantom shadow that slipped past at some twenty knots under the eerie light of the moon. Once more they cried out, shouted, raised their arms toward the skies. Once more they were ignored.

It was day again. Again, they had pumped up and emptied the raft. The pump had become harder and harder to work. Catherine was now unable to use it for very long. Lucien was only able to make it work by falling down on top of it, pushing down the handle with his chest.

Having finished that chore, each of them again took refuge in their "gray soup," made uneasy by dreams and sufferings. But Lucien soon roused Catherine. He stared at her with the lost look of someone who could not escape from a dream that clung like a second skin. He asked, "Where is the fresh bread?" She couldn't put up with that sort of intrusion. She replied harshly, "Stop that nonsense. There is no fresh bread."

"Yes, there is. The submarine just brought some."

It had been a yellow submarine. It had surfaced beside the raft. Some officers had climbed onto the deck. What they had offered had not been fresh bread but hot croissants. Then they had apologized. They were in a hurry, didn't have time to do more. They had to dive again quickly. Lucien had thanked them. The submarine had disappeared.

"There is no fresh bread," Catherine repeated.

"All right," Lucien said. "I must have been mistaken, it doesn't matter."

He withdrew within himself and was again with Tony, the boy who lived in a minibus equipped like a caravan, with a gas oven, a refrigerator so stuffed with food that when one opened it fruits fell out and rolled on the ground. Lucien smiled at the sight of such abundance. Tony asked him, "Would you like some coffee, a nice cup of coffee?"

Catherine was with her friends in Paris to whom she was telling her story: saved on the twelfth day. She was there alone; once again Lucien did not enter the picture. Did that mean that Lucien was going to die?

They drank three, four swallows, then five, then six. What was the point of rationing themselves from now on?

What was the point of hoarding that pathetic supply of water that kept them alive without ever allowing them to break away from their destiny? They might as well not talk about it, drink their fill and let themselves slip into unconsciousness, forget their burns, their kidney pains, their stupor. Lucien talked about suicide but it wasn't very convincing. Catherine could not take his decision seriously.

Another drink. Perhaps draining the jerry can was the best way to end everything. It was a bit like leaping off a cliff. Once underway, nothing could save you.

Yet Lucien did not want to drink. Now that he had chosen to die he found it pointless to help use up the supply. Let Catherine have it. It gave her one more chance to survive.

"Just as you like," replied Catherine, "but I think your attitude is absurd."

He hesitated, realized at last that she was no doubt right, and swallowed a few more mouthfuls.

From then on, they had less than ten liters left. Unless it rained again, the die was cast. They would last another forty-eight to seventy-two hours at the most. Beyond that . . . well, it was hardly worth thinking about.

"It would be a shame," thought Catherine. "I had so many things to do in my life, things that will never be done."

8. Fourth Conversation: Hunger

Catherine: Having come this far, it would be impossible to go further without mentioning something we have avoided until now. It's just too bad for the softhearted. If we are to speak the real truth about this matter, certain forms of prudery, or rather certain social hypocrisies, are irrelevant. We must go all the way.

But first of all we should make clear again what our situation was on Friday the 22nd, the eighth day, difficult though it may be to set precise boundaries on our thoughts. When I say Friday the 22nd, I'm no doubt simplifying. What each of us in our little corner imagined that day, we had no doubt already thought about, more or less confusedly, the day before or even the day before that, if not earlier.

In any case, one thing is certain. That kind of thinking did not arise in our minds all at once. We reached that point after many meanderings. We must have thought about similar things earlier, but without attaching any importance to them since they were speculations of no immediate concern. Then, little by little, those so-called speculations asserted themselves as actual plans. We began to consider them with complacency, until the thing itself was so deeply embedded within us that it became one of the central themes of our thoughts and dreams.

And when I speak of us, I mean each of us taken separately. Anyone will readily understand that in our situation an intellectual or sentimental community was no longer possible. If I spoke, a few moments ago, of the reality of the situation I was referring to this fact. Aside from rare instances of rudimentary exchanges, each of us was enclosed inside his own world. There was no question of allowing the other access to it. Our solidarity, or what was left of it, was only revived for a moment when our survival depended on carrying out some essential act together, like pumping up the raft, bailing, sharing the water, or signaling an eventual rescuer.

I say "we." That's perhaps unfair. After all, Lucien may not have lived through that period in the same way. Yet I think he did. Several indications led me to think that we had reached the same point. The moment when we thought that if the other were to die, it would of course be in many ways dreadfully lonely, but it would also mean the possibility of nourishment again, of regaining strength, of saving one's skin.

"In other words, by Friday each of you seriously thought of nourishing yourself with the remains of whoever should die first?"

Lucien: As far as I'm concerned, that's right. I had decided that I must survive. I'd always had the impression that Catherine was much weaker than I. Therefore, it was logical that she should die first.

Catherine: I held the opposite point of view. I thought, "Lucien is very low. He won't last long or he'll

Fourth Conversation: Hunger

give in to one of those suicidal urges that overtake him now and then." I saw his body, his face changing. I read death in his look.

"Had you, in any way, referred openly to such plans?"

Catherine: Yes, but in a very hypocritical way, each of us saying that he wouldn't eat the other if he died. That was a way of admitting that we wanted to do it. Yet I admit I was quite surprised when I learned that Lucien held the same ideas on the subject. I didn't think he was capable of it.

Lucien: I was also surprised. I reproached myself. I thought, "How can I think of eating her when she would never dream of doing that to me." And then at other times I was ashamed. I thought of the days we had spent together. I realized that even after she died I would be unable to grapple with her. And yet, little by little, the obsession with food regained the upper hand.

Catherine: I, too, felt ashamed, told myself I couldn't do it and at the same time I thought, "There's no reason why it should be wrong."

"So there came a time when you began to watch each other, to say to yourselves that the death of one of you could mean a kind of salvation for the other?"

Catherine: Yes. And I began to see what had to be done very quickly. We had an absurd little knife, no doubt thought up by some survival expert sitting in his comfy armchair, a knife designed to do everything. There was a bottle opener at one end, as if we'd have brought bottles of soda or mineral water. In short, a knife that's good for nothing.

I saw myself trying to carve with that knife. What piece should I choose? I felt like a cutlet, but that was probably very difficult to remove. What then? The buttock seemed unappetizing. I wondered where steaks were in a human being. Wouldn't it be simplest to cut a piece from the thigh? I was also tempted to suck out all the blood. Could it be done? How should I go about it? I thought about that sort of problem for hours.

From time to time, however, I would come up against a stone wall. First of all, education, accepted notions, the revulsion of a civilized person, although on the eighth day not much was left of the great social taboos. At other times they were more down-to-earth obstacles. For example, I saw myself being saved by a boat with a half-carved body next to me. I was being questioned, I was thrown into prison for the rest of my life. Then I thought of rigor mortis. I pictured this great body, stiff as a board, stretched across the raft, pushing it out of shape. I wouldn't be able to do anything with it, to drink its blood, to carve that monolith. I'd have to live with it, to let it have the lion's share of room, to step over it continually. Little by little that cadaver consumed everything—an enormous statue too heavy for me to throw into the sea.

But that nightmare lasted only a short while. I soon convinced myself of its ridiculousness. If it came to that, I wouldn't be the first to be in such a predicament. Others had eaten their dead companions in order to survive. We had even heard of people killing one another to provide food. After all, I hadn't reached that point. In all frankness, although it had occurred to me to wish for Lucien's death, I never for a second dreamt of helping to bring it about. The thought never crossed my mind, not even an eventuality that one rejects. It was always, "If he dies," or "He's sure to die." And, starting at that point, the dream unfolded itself more and more precisely as time went by. I

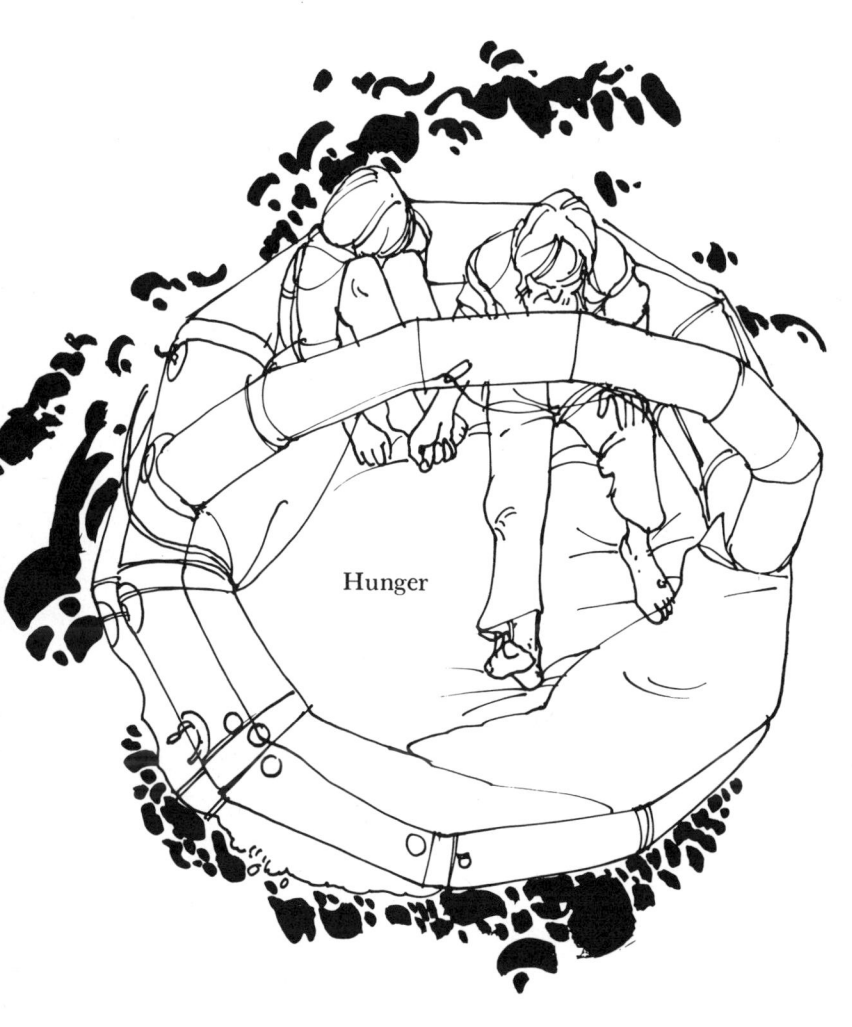

took pleasure in the evocation, in the meticulous inventory of the gestures destined to ransom my life.

"What had become of that noble decision not to let yourselves dwell on the thought of food?"

Catherine: It was still there. Please understand. I had always forbidden myself to dream of nonexistent food, to allow myself to succumb to the impossible. But obviously, if by some miracle food had appeared in the raft, I would not have complained about it.

Now, the moment I thought that Lucien was about to die, and that it was not only possible but certain that I would eat his flesh after his death, the food appeared, so to speak, or was about to appear. The living Lucien had ceased, in a way, to be a companion, a human presence—although in another way, of course, he still was—and became, instead, the possibility of food that could be transformed into reality at any moment. So it became natural to picture the new situation and to hope it would happen.

Lucien: I thought along pretty much the same lines, except that I also dreamt about other kinds of food.

When I was dozing, I went to a little restaurant in Beaulieu where we used to go. In the middle of the restaurant there was a table piled with hors d'oeuvres. I helped myself and ate and ate. And then, when I was back in the raft, I saw Catherine opposite me. I had the impression she was growing weaker by the hour. She scarcely moved. She had told me that she felt stiffer all the time, that it was harder and harder for her to make the slightest move. It was obviously the end for her. So why turn down that chance to survive?

I also wondered how to go about carving her up, what piece I should remove first. I thought of her liver. I'm very

Fourth Conversation: Hunger

fond of liver. I had asked her, "Catherine, is your liver healthy?" She had replied yes while looking at me in surprise. She even said, "Why do you ask that question?" I told her, "Because I think one needs a healthy liver in order to survive."

Catherine: I remember. I immediately gathered that he was planning to eat my liver if I died. I think I was sure then that he too was considering feeding on my corpse. Yet I was not unduly impressed by the discovery. I felt like telling him: naive Lucien, you are sadly mistaken if you think I will be the first to die.

"And yet you continued to reassure each other that if one of you died the other would not eat him."

Lucien: Yes, it was one of the basic themes of our rare conversations. From time to time we would talk about death, or rather about our ever more slim chance of survival. We were asking each other how we felt, if we thought we could last much longer, without realizing too clearly that each of us understood perfectly why the other asked those questions. And then that led immediately to our chief preoccupation: "Do you think you would be capable of eating me if I died?" There followed vehement denials which neither one believed, yet which didn't prevent our raising the subject again at the first opportunity. It was as though we wanted to reassure and unmask each other at the same time. We had to know where we stood.

I remember trying to trap Catherine. I said to her, "After all, if I died first and it was your only chance to survive, I don't see why you shouldn't eat me. It wouldn't make much difference to me, once I was dead."

I don't remember anymore exactly what she replied. She must have reassured me that she would not be capable of it, which somewhat set my mind at rest. In effect, I was always trying to convince myself, against all the evidence, that even if I had decided to eat Catherine's flesh when the occasion arose, she certainly wouldn't behave in the same way toward me. She was thinking of it, perhaps, yet when the time came, she wouldn't dare do it. And at the same time I said to myself that the problem wouldn't arise for her in any case, since it was quite obvious that if anyone were to die it could only be she.

Catherine: I had no illusions. I knew very well that if I died first as the result of some unforeseen accident, Lucien would eat me.

Lucien: And yet, perhaps I'd have hesitated. I had a rather dreadful dream several times. I was landing on some shore with the half cut-up body of Catherine in the raft. Luckily the beach was deserted, yet I was afraid someone might come at any moment. I wanted to run, to run as fast as my legs could go to avoid being discovered next to that mutilated corpse. Yet if I abandoned it there, there would be an inquiry about it. It would lead to me in the end. I'd have to live the life of a hunted man. So I decided to get rid of it. I cut up the rest of the corpse and threw the pieces into the sea. Then I began destroying the raft, strip by strip, because of bloodstains. Once having done that, I headed for the interior. I dared not let myself be seen. I hid in the forest. I was going to feed myself by stealing food from the farms at night. I dug up potatoes from the fields. I went inside henroosts, making off with eggs, wringing the necks of chickens.

"And what happened when you woke up and saw Catherine facing you, still alive?"

Lucien: First of all, I was relieved. Compared to that nightmare, the situation in the raft seemed, if not desir-

Fourth Conversation: Hunger

able, at least tolerable. Nothing irreversible had yet taken place and there was still hope of getting out alive. And then I began to wonder whether that dream were really a nightmare. It had several ghastly aspects, of course. Yet the potatoes, the eggs, the chickens, renewed strength, one's fill of water—these things were far from unpleasant.

The more I thought about it, the more I was obsessed with the idea of food. Only a few moments before I'd been telling myself: you know quite well it's impossible, that no matter what happens you could never cut up Catherine's body, eat her flesh. And yet, there it was. The temptation was coming back so strongly that there was no question of resisting it.

I saw the problem in very simple terms. As soon as she were dead there would be only two alternatives—to eat her or to die.

Yet wasn't I a bit hasty to reduce it to those two alternatives? More and more often, especially during Saturday evening and night, I wondered whether things would ever happen that way. When the moment came, would I still have any strength left to make the slightest effort? Suppose I died first? I was beginning to consider that possibility. In that case, why not hasten the outcome? Wasn't suicide the only real solution?

I pictured myself dead. I had become something else. Perhaps a seagull. Yes, why not a seagull? Or perhaps another animal. A cow, for example, belly-deep in rich grass, endlessly contemplating an unchanging landscape through vacant eyes. That prospect made me freeze with terror. No, I couldn't run that risk. There was only one solution. To live, whatever the cost.

Catherine: It wasn't death I was afraid of, I mean death as the beginning of something else. For me, death was the end of everything—in other words, the final, absolute end of Catherine. That was what I rebelled against, losing my life, now, so young, while there were still so many things to experience. And losing it that way, almost alone in that raft. No one would ever know what had happened. There was no way to leave my story behind. If only I could have sent a message, at least to say what my life had been like to the last day.

I had the impression of being walled-in. Yet, at the same time, I felt my situation was worse. In the case of someone walled-in, there would be others to know about it. The victim could always hope that at the last moment someone would take pity, would come to rescue him or that he might manage in a final outburst to topple the stones that separated him from the world.

III. THE OUTCOME

9. Land

The weather was fine that morning. There was a clear sky, a calm sea, and a very light northerly breeze. "It's going to be warm," thought Catherine. "If only I could undress and sleep in the sun!" It had become an obsession for her. To strip naked, to rid herself of those sopping wet clothes, saturated with salt, and stretch out on warm sand in the sun in the hollow of a dune, protected from the wind, to ease the pain caused by the many burns on her body.

As he did each morning and several times during the day, Lucien, before pumping up and bailing out, was inspecting the horizon. To do this, he had to stand up, because a good part of the forward view was hidden by the sail improvised from the remnants of the tent. But standing up was an exhausting and painful exercise. That was why he did not give in more often to his constant desire to see whether anything was happening on that part of the sea not visible to him from his customary position. He often contented himself by craning his neck a bit above the pneumatic rings to increase his field of vision. Catherine did the same thing on the other side. It was enough to establish whether anything was happening.

Instead of confining himself that time to his usual sweeping glance punctuated by a discouraged shrug of his shoulders, Lucien remained motionless, apparently contemplating something with sustained attention.

"What are you looking at?" asked Catherine.

"It's very strange, look."

She did not want to move. Her muscles felt as stiff as if she had tetanus. It was the first time she had experienced that painful sensation. It worried her. In addition, whenever she moved, the water permanently trapped in her oilskin pants would move, chilling her and irritating her burns.

She glanced outside the raft but saw nothing. "You really want me to look? You can't describe it to me?"

"I'd like to have your opinion. There, just in front of us, directly on our course, a layer of cumulus, like a crown on top of something. It doesn't make sense in this clear weather and northerly breeze, unless there's land under those clouds . . ."

Land! Another one of poor Lucien's hallucinations, like the chocolate mousse, the shower, the couscous, the yellow submarine. Besides, what land? Spain? It was much farther away. The Balearics? That would be more likely, although . . . Once again she tried to crane her neck outside the raft. She actually caught sight of a patch of clouds.

"I assure you," said Lucien, "it can't be anything but land, probably one of the Balearic Islands."

She refused to play the game. She knew that the moment she let herself think they really were near land, a disappointment would prove unbearable. She needed all her strength to hold on until the twelfth day. She couldn't afford to waste it on pointless excitement.

She looked at the white plastic jerry can beside her that held the key to the future. The evening before, they had drunk a great deal, far too much—the worst mistake they had made since abandoning *Njord*. It was essential now to get a grip on themselves and resume their prudent rhythm of survival. There must have been only five or six liters of water left, barely enough to live on for forty-eight hours. In two and a half days they had consumed nearly fifteen liters—insane behavior that had not even brought them satisfaction for, despite their lapse of self-discipline, they had never really quenched their thirst. It was not a burning thirst that coats one's mouth and stops one's throat, but a chronic, insidious, perpetually frustrating thirst.

And thirst was only part of it! They had to contend with fatigue, burns, stiffness, stabbing pains in the heels, and that hunger which had nothing in common with what people in the everyday world meant when they said, "I'm hungry," as they sat down to eat. It was no longer a need, it was a visceral cry. These were the things that gave rise to an inescapable uneasiness, an incapacity to react even in the face of a fabulous possibility: land.

But any determination not to waste energy on needless motion was no match for the fascination created by Lucien's saying, "Land, I tell you, it's land."

Catherine, gripping the half-deflated arch, raised herself on legs that she had thought were permanently bent.

She was standing at last. She saw now the crest of clouds that contrasted strangely with the clearness of the day. The northerly breeze had freshened a bit. The raft was moving toward the pile of clouds that was, perhaps, only a mirage. The flexible bottom of the raft conveyed the rhythm of a slight ripple. That sound made her think of the ripple one sees on a pond from a boat anchored among the reeds when the first morning breeze comes up and sweeps away the pale colors from the smooth water.

While those images were unfolding, something knotted itself inside Catherine's chest. Iridescent bubbles swollen with memories came bursting to the surface of her consciousness. God! So much to live and relive, so many pleasures to experience, emotions to master! Could it really be that nothing more would ever happen?

"I seem to make out something, something dark below the clouds," said Lucien. With his left index finger he pointed to that spot on the horizon. His right hand, resting on the arch, trembled a little. Catherine gazed intently at the grayish mass of clouds that seemed to melt into the sea, changing the sea's color. No, she saw nothing to alter her scepticism. Perhaps Lucien was right. Perhaps that opaque mass really did indicate the presence of land, but there was nothing to confirm it. The sea was still as hopelessly deserted as ever—not a bird, not a fish, no flotsam, not the slightest sign of any form of life.

Slowly, cautiously, she resumed her position near the valve of the upper ring so she would be able to pump without having to move again. She behaved mechanically, attaching the pump to the valve, pumping, stroke by stroke, pause after pause. The arch stiffened again, the folds disappeared from the ring. Lucien would see to the second valve. She could not reach it without having to move. He would also see to the bailing, which was becoming ever more painful a task. The night before, they

had let things slide a bit in the face of having to make such a painful effort. The result was almost four centimeters of water in the bottom of the raft, the limit beyond which it was impossible to use the pump.

Still standing, Lucien continued to scan the horizon. He could not tear himself away from the contemplation of those clouds. For nearly two hours he had been studying their shapes, straining to uncover permanent features that might prove to be more than mist, believing he had found some, then, after a moment, having to accept the truth. The peak, the cliff, the rounded outcropping he had been so sure of, dissolved slowly under pressure from the light breeze, were drawn out into wisps, and took shape as other three-dimensional phantoms.

He had interrupted his vigil for a few moments only to look after the raft, to finish pumping up and bailing out. Catherine suggested he spare himself, sit down, wait. Her advice fell on deaf ears. He could sense land, feel its presence. Nothing could have torn him from his vigil. Suddenly, toward 10:00 A.M., he cried out, "This time I'm sure. I see it, I see an island. Land! It's really there! The wind is pushing us there! We're saved!"

Catherine leapt to her feet, amazed that she could still move so quickly. And she, in her turn, was confronted by the stupendous fact of land. Of course, the picture was not yet complete with the features of a coastline, yellow rocks, the bright patch of a beach, and the thousand signs that revealed human presence. All she saw was a bluish mass whose contours were smudged by distance, a dark band across which trailed shreds of fog.

Was it Majorca, Minorca, or Ibiza? Ibiza was most unlikely. How could they have reached it without first having sighted Majorca? Most likely it was either Majorca or Minorca. What difference did it make? The main thing was to land somewhere.

"It happened all of a sudden," said Lucien. "I was still trying to see something in the clouds when they suddenly cleared up."

How far away were they? Probably about twenty miles. That was a long way. Everything would depend on the wind. With a bit of luck they could cover that distance in fifteen hours, which would bring them ashore in the middle of the night.

Landing in darkness was not ideal. It would be better to go slower and come ashore in daylight. In any case, they had no choice. They had virtually no way to maneuver. All they could do was wait until the sea had taken them where it would.

Making a landing was not the only solution possible. The chance of encountering a boat was always greater close to shore. And it was no longer just a matter of commercial vessels, which until then had treated them with such sovereign indifference. Fishing boats or pleasure crafts were bound to cruise where they were. People moving very slowly and, like themselves, very nearly at a level with the water. Why not hope to hail one of those stout boats whose owner had just pulled up his lobster pots? To sit on the bleached wooden deck, amid the traps, and share his cheese, his bread, his thick wine while the engine took them gently toward the harbor?

True, it was Sunday, which was probably a handicap. Perhaps Spanish fishermen did not work on Sunday. On the other hand, it was a perfect day for pleasure boating. But, wasn't it already a bit late in the season?

"What about a swim?" asked Catherine. It was

11:00 A.M., the sun was hot. It was the first time since they had taken to the raft that the temperature had been so pleasant.

"Do you think it's a good idea?" asked Lucien. "We're so weak." Catherine insisted, "I feel like it. I have to shed this sticky armor I'm trapped in. It would do me good to be naked, and to take a dip in the sea. I have a feeling it would soothe my burns."

The breeze was still light. Lucien acquiesced. But they had to undress, and that was so painful they were on the point of giving up several times. The last time they had undressed was five days earlier on Tuesday the 19th. It had taken a considerable effort then to remove their clothes, but nothing compared to what it now entailed.

Their bodies were in much worse condition. There was hardly a place where the skin was not burned, and those burns were particularly deep at the joints, the shoulder blades, buttocks, thighs, and wherever friction had aggravated the action of salt water. They were also appalled by how thin they were. "Our legs were like matchsticks," Catherine said later. "It was really frightening."

They let themselves slide into the water but this time dared not let go the lifeline that ran round the raft. They were too tired to try to swim. Nevertheless, they experienced a feeling of great relaxation. Catherine thought she felt her muscles loosen.

The rising of the wind forced them to cut short their pleasure. The raft began to move. Had they let go, they wouldn't have had the slightest chance of catching it. It was time to climb back aboard, and that was no easy matter. They would learn in a few moments whether the bath had been a terrible mistake. What if they were unable to climb back? In the end they climbed aboard, but only after pushing their strength to the limit. They had hoped to sun themselves a while but weren't given the chance. They were shivering from wind and weakness. They would have to dress right away, irritating their burns. Once dressed, the feeling of being cold continued. Their muscles contracted again and they had cramps. It would take them a long time to warm up.

The wind had reached force four and the raft was making for land at better than a knot. The sea was blue, alive, cheerful. They had left the fourth dimension. The universe had regained its familiar aspect. From now on they were going somewhere, were heading for land, quite visible now that the clouds had blown away. Lucien thought he recognized Majorca. The rocky spur before them would be the northern extremity of the island. Behind it, on the northeast coast, would be the big bay of Pollensa.

Lucien described the terrain, which he knew. If the wind held, they would probably leave the rocky point to starboard and come ashore in the southern part of the bay, on a beach with the small town not too far away. What would happen after that? They spoke about it, they dreamt of it.

Everything might work out well. They might soon meet friendly people who would see their plight and come to their aid. They would have to offer some explanation. Catherine did not know Spanish, but Lucien knew the fundamentals. He tried to remember them, to construct a few sentences. He imagined a dialogue: "Who are you?" "We're shipwrecked." "Where do you come from?" "From over there, from the sea."

Given their condition, they might be taken for vagabonds and brought to the police. "We'll surely be able to explain," said Lucien. "First of all, we have our papers and money." It was true. Their money and their papers! They had forgotten about them. They were still in the canvas pocket where they had put them the day they had tried to bail with the leather pouch. What had become of the leather pouch? Lost, no doubt, during a capsize. They hadn't noticed.

They spread the bank notes and the identity cards on their legs to let them dry away from the wind. Those bank notes! Catherine stared at them. There were almost 5,000 French francs. Those pieces of paper were beginning to become precious again. Yet a few hours ago she would gladly have exchanged them all for one apple.

Catherine thought of what would happen once they had landed. She pictured herself walking along the shore, reaching the town. There would be difficulties at first. She would be questioned, she would have trouble explaining, but everything would be all right. She would go to a market, buy oranges and lemons and a large glass of fruit juice. What had become of Lucien? She remembered suddenly that she had left him on the beach, near the raft, unable to take a step. She had gone in search of help. Now no one wanted to believe her, to help her. When she returned it was too late, Lucien was dead.

Or then it was she who could not walk. Lucien had left, taking the water supply with him. He had said to her, "Wait for me, I'll come back with whatever is needed to carry you." How long had he been gone?

The stiffness of her muscles, from which she had suffered since morning, had been growing worse. It was no longer just her legs that were affected but also her arms. She asked, "You don't think we'll have a great deal of trouble getting help?" No, Lucien was confident. He felt he was still strong enough to find help. Besides, they avoided dwelling too much on the problems that might arise after landing. After all, the main thing was to land. They would worry about the rest afterward.

One thing was sure. They would take a holiday, a real middle-class holiday in a family pension. They would write to confirm their booking and then they would enjoy life. "And then we'll be married," declared Lucien. "After what we've been through together, we couldn't possibly separate." "Agreed," said Catherine, "we'll be married." It was not the first time they had made such plans, but for two days those plans had been quite forgotten while each had viewed the other as nothing more than potential food. Now the proximity of land had rid them of that obsession.

The raft continued to move in the right direction, the weather was still as fair, the sun as warm. The distance separating them from the island had decreased noticeably. They must have overestimated it in the first place, because toward 4:00 p.m. the island seemed only a dozen miles away. If the wind held, they should arrive during the night. But wouldn't the wind die at sunset? And which way would it blow in the morning? No, things couldn't go wrong this time.

Catherine, remembering suddenly her dream of the twelfth day, was seized with despair. The dream might mean that their landing was not to take place, or at least not for another forty-eight hours. Was it conceivable that they would stay close to the island another forty-eight hours without being able to land on it? She refused to accept such a possibility. For one thing, they did not have enough water to last another two days. And what about the dream . . . ? It was all right to believe in something like that when there was nothing else to cling to, but now that land was in sight surely there was no longer any need to think about it.

Toward 5:00 P.M., however, the facts had to be faced. Not only had the wind fallen but a current was taking them a little to the southeast, which might well carry the raft on a course parallel to the coast. Yet the wind, or what was left of it, did not seem to have changed direction.

"It's probably a current that forms in the channel between Majorca and Minorca," said Lucien.

The presence of the current was extremely disquieting. If it were constant it might, in the absence of wind, carry the raft southeast throughout the whole night. And how could they prevent it?

A big boat was heading in the opposite direction between them and the land. It was going to pass too far away for them to be able to attract the attention of its crew. "If only I had a mirror," thought Lucien, "I could use the sun to signal them."

He was furious to think that such an encounter was at the whim of wind or current. He grumbled. "We must do something, we can't let ourselves get further away from that island without doing something. If only we hadn't lost the oars!"

If only we had this, if only that had not happened. That had been the leitmotiv for ten days. A means of

propulsion, to find a means of propulsion for approaching the shore, for covering the dozen miles that remained. To have come from so far away, to have overcome so many obstacles only to find themselves frustrated, without the slightest recourse, within grasp of their objective! If only they were immobilized, but there was that wretched current! How could they move in the right direction? Row with their hands? Why not? Every meter gained might prove precious afterward. Lucien succeeded in convincing Catherine to try. For several minutes they squandered the small remnant of their strength without any measurable result. It was then that Lucien decided to risk everything. He was going to go over the side and tow the raft toward shore. He felt he was up to it. He felt he must do anything rather than wait for something to happen, something which might well prove to be fatal like another tramontane that would hurl them beyond the islands forever, toward the south, into oblivion.

Catherine tried in vain to dissuade him. Using the piece of cord, he rigged a painter, secured one end to the raft, wrapped the other round his body and threw himself in the water fully clothed. For nearly ten minutes he swam toward land, toward life at the risk of his own, until he was out of breath. He stopped, utterly spent, after having advanced perhaps twenty meters toward his goal. Fortunately, his life jacket buoyed his head above water, for he could no longer move at all.

Catherine hauled him back slowly by pulling in the towline. She knew she couldn't pull him aboard so long as he remained listless, his eyes rolling, his breath coming in gasps. She could do nothing for him if he didn't take a hand in his own rescue. But he soon began to stir again, holding out his right hand while he gripped the upper rubber ring with his left. If he hadn't been such a pitiful sight she would have exploded with anger: he had made things worse for nothing, as though things were not bad enough already.

Painfully getting down on her knees, offering her weak arm as a purchase, she despaired of ever seeing him, weighed down by soaked clothing, aboard the raft again.

Lucien, however, did not give up. He let go and fell back for a moment into the sea above his chest. He began to breathe more easily and then hoisted himself up, resting his elbows on the upper ring and, in the same movement, putting his feet over the lifeline. Catherine gripped the seat of his oilskins and found the strength to help him flop inside the raft.

He was stretched out at her feet in the bottom of the raft, his arms forward, his belly still on the rubber gunwale, and his legs dangling over the side. It took him several minutes to recover. He was trembling and mumbled incoherently. He felt as though his lungs had turned to stone as he couldn't fill them with enough air.

After a while he was able to move. With Catherine's help he resumed his customary position, half seated, half lying. The stone in his chest continued to oppress him. His kidneys were frozen. He felt that the cold was part of his being, that nothing would ever rid him of it.

The wind had died completely. The sea's only motion was an imperceptible swell. The sun was setting to the right of the island. They were still slightly to the northeast of the nearest spit of land. It was 6:30 P.M. as Catherine waited for Lucien to recover from his exertions. The current that had just then caused them so much anxiety seemed to have disappeared. The bearings she was able to take against the land indicated that the raft was not moving except to bob up and down in the splendor of that almost-summer evening. The only thing they could hope for now was a favorable northeasterly breeze. An east wind would not be enough to ensure their coming ashore. Given their position, it would take them north of the island, which would mean losing all hope. And there was one more requisite. The wind must rise with the sun and be strong enough to let them complete the last leg of their journey before lack of water and exhaustion left them helpless when they landed.

Toward 7:00 P.M., Lucien came to life. "I feel better," he declared. "I'm less cold, I feel revived. Also, I've thought of something. We're going to stream the sea anchor. Imagine if an offshore breeze comes up and pushes us back several miles out to sea . . ."

A little later, when it was almost dark, they witnessed an impressive sight. They had just completed again the exhausting task of pumping air and emptying water. They had each taken two swallows of water and had settled themselves for what was to be their tenth night of survival. At the very moment when the sun set, the full moon rose. Suddenly the whole island was illuminated from behind by the flaring of the sun and from in front by the full force of the moon's incandescence. There was something remarkable about that confrontation. They had the impression that the tiny rubber raft was in the exact center of an immense balance beam at the ends of which, for an instant, two heavenly bodies were poised in equilibrium. But the sun soon disappeared, and with it one of the arms of the balance beam sank below the horizon while the moon rose and the raft remained motionless in the deep silence of a perfectly calm night.

The land was either too distant or too deserted at that spot for them to be able to distinguish any lights. The darkened mass of the island was barely discernible anymore. A brilliantly lighted liner moved along the coast. They watched it without making a move, convinced of the futility of any effort. "Tomorrow," said Catherine, "there'll be a northeaster. It can't be otherwise."

She closed her eyes and prayed, without a great deal of faith. What she invoked was a sort of father of miracles, her personal God, the one whom, in other times, she called her luck. "You know me, you accept me as I am, you have often given me your helping hand, pulled me through bad scrapes. Was it to abandon me here that you brought me within reach of safety?"

Lucien was sleeping now and was dreaming. He had left the raft and had realized his desire to reach the shore by swimming. He had succeeded without too great an effort as land was much closer than he had suspected. He climbed from the sea, walked along the strand, followed a dirt path, ran toward shuttered houses with closed doors. He must get help. Catherine was waiting there, at sea. He must go to her rescue. But no one answered his appeals. On his way he stole food from bags he found along the path, from stalls that no one was minding and which had remained open despite the lateness of the hour, and then took a steep trail that led to the edge of a cove. Several small boats were rocking in the moonlight. He made for one, climbed aboard, hoisted the lateen sail, and weighed anchor. But nothing happened, there was no wind and there was neither engine nor oars. He had to rejoin Catherine who had remained at sea on the raft. But where? How would he find her? Had he lost her for good? How would he bring her the fruit he had gathered for her?

The cold woke him. A fine rain was falling, too fine to furnish drinking water. He tried licking the tent canvas to ease his thirst, but the taste was so awful he gave up. Why was it raining? Had the wind shifted? The air seemed quite still and the sea was so calm and the raft so undisturbed by any force whatsoever that the sea anchor's line was not even taut. Besides, the rain did not last. After a quarter of an hour the sky cleared and the stars reappeared. A boat passed along the coast, then a second, then a third. He wondered if it was worth waking up Catherine, who seemed to be asleep, but decided to wait until day.

10. Fifth Conversation: Fog

Catherine: When the sun came up Monday morning, it was marvelous. The island was still there, perfectly visible although a little shrouded by fog. Judging by the bearings we had taken, we had hardly moved during the night. The current that we had detected for a while had not outlived the wind. It was a beautiful hot sunny day. From the first hour of daylight we knew it was going to be particularly hot. It was true that we were only at the end of September—still the sunny season in the Balearics. But we had had so little luck with the weather since our departure.

There was no swell, just a gentle ripple similar to the one which had worried me briefly last night. And yet the air was quite still. There wasn't a breath anywhere. About 8:00 A.M., the sea began to show patches of white in the distance. We followed the birth of that wind with apprehension. Lucien said, "Given this sort of weather, the wind is almost sure to come from the same direction as yesterday." I had trouble overcoming my anxiety, wondering if we would be blessed by a northeast wind that would take us toward land. What if it were a sea breeze that brought us closer only to be carried away in the afternoon by a land breeze?

I pictured us, come evening, being returned to the same spot after having spent the whole day in a pointless coming and going. There was the sea anchor, of course. It would allow us to counteract our drift, to retain our present position in the hope that on the following day a sea breeze would rise once again that would let us make a few more miles in the right direction. But, we had to arrive that very day. We couldn't remain suspended between life and death any longer. In any case, the next day, Tuesday, the twelfth day, we would have no water left and how could we do anything once ashore if we were hopelessly weak?

Lucien: I must add that I was increasingly worried about the pump. While waiting for the wind that morning, I had used it as usual to pump up and bail out. Bailing out had become so difficult now that I was afraid I wouldn't be able to do it. It's true that I must have had less and less strength. But that wasn't the only problem. The pump was reaching the point where it could no longer be used and there was no question of repairing it. It was designed in such a way that it couldn't be dismantled.

Also, Catherine said that she didn't share my optimism about the direction of the coming wind. Well, I wasn't so sure either. By appearing confident I was trying above all to reassure both of us. When, about 8:30 P.M., a northeast breeze rose and I again felt the raft move toward shore, I gave an enormous sigh of relief.

Catherine: Yes, we were incredibly happy. There were still a dozen miles or so to be covered and the breeze was blowing in the right direction. It was quite strong, just what we needed to make us feel that we were getting some-

Fifth Conversation: Fog

where at last. And it was clearly not just a sea breeze. Sea breezes come up more circumspectly. If all went well, we would arrive late in the afternoon, before night fell. That's how favorable the circumstances appeared to us. And then, it was a Monday which increased the odds of our running into a fishing boat.

Our first concern was to rig the strips of cloth that we used as a sail to try to increase our speed. We used up practically the whole ball of string in the process. In my excitement I was hardly aware of my burns. But I was very handicapped in my movements by the growing stiffness of my muscles.

Having patched up the sail, we each had three swallows of water, settled ourselves in our usual positions and took out our bank notes and identity cards, which were wet again, to dry them.

"What were you thinking then? Were you quite confident? I mean, as confident as one could be? Did you say to yourselves, 'This time we're saved,' or did you still have some doubts?"

Lucien: I was confident. There was apparently no reason to be worried. The wind had risen in the morning. It was blowing steadily. It seemed to have dug in, so to speak. It was probably going to freshen during the day without, however, exceeding force four. It seemed to me that only one unpleasant thing could happen to us. The wind might drop about 5:00 P.M., as it had the night before. But at that point, even if we hadn't arrived, we would be very close, one or two miles away at the most, and therefore inside the bay. Besides, I felt it would really be rotten luck if, having come this close to shore, we didn't run across a fisherman to help us. Finally, the fact that we would, in all likelihood, come ashore in the bay was a great piece of luck. We would have no trouble making a landing there.

Catherine: I was also confident. I thought we'd make it that time. If we didn't manage to reach land that day or, at the very latest, the following morning, the twelfth day, our chances of survival were so slim it was better not to talk about it.

"Did Lucien know about your dream of the twelfth day?"

Catherine: No, it was my secret hope. I had saved it for myself. I only told him about it later. We'll come back to that.

"How did you feel physically?"

Catherine: I was at the end of my strength. Had it not been for my desperate desire to be out of that mess within a few hours, I think I would have collapsed. My burns were causing me a great deal of suffering and my cramps were more and more painful. I could hardly move anymore.

Yet I felt that, once the moment came, I'd still find enough strength to get out of the raft, to set foot on land, to walk. And the sun's warmth was increasing by the hour. It was marvelously pleasant despite my many pains. I felt that if I were really able to warm up, my legs, my arms, and my back would recover at least some of their mobility.

Lucien: I felt I was in relatively good shape. Very tired, of course, but not exhausted. The day before, after having foolishly tried to tow the raft, I had been very afraid. I thought I had reached a threshold, and wondered if I would ever recover. It had taken me a long time to warm up and to breathe freely again. Yet it reassured me that after only one night of rest the traces of that stupid and taxing effort had disappeared. Of course I also had

Fifth Conversation: Fog   157

my complaints, especially those wretched burns, but I thought that there was no reason why I shouldn't last until the end.

I felt a great deal better than I had during the two or three days before our discovery of land. That period, when we were considering cannibalism in our separate corners, had become almost as strange to me, seemed as mad to me as the first period, that of the storm.

And then an extraordinary thing happened about 10:00 A.M., when everything seemed to be going well. The island suddenly disappeared. It didn't take more than a few seconds. The fog, that until then had given the island a sort of bluish halo, suddenly thickened until the island was completely hidden. It was still a fine day, the sky was cloudless, the sun warm, the horizon clear, the sea still a deep blue, that matchless blue of the best Mediterranean days. But there was no longer any land or at least any visible land. There was not even, as there had been the day before, a crest of clouds that had drawn my attention. It was as if someone had rubbed out the landscape or rather had hidden it behind a luminescent screen, like frosted glass.

From that moment on, there was no way to tell where the northeast point of land was or the bay. Only the direction of the wind gave some indication. But what could we rely on if the wind changed direction? How could we tell if it changed? The fog that had engulfed the point we had been making for was spreading so rapidly into the sky, we couldn't make out its exact limits. That left the sun, but the more it approached its zenith the less precise were its indications.

Well, there was nothing so unusual about those conditions, it's true. Fog caused by heat is a well-known phenomenon. And why should the wind take advantage of the island's disappearance to make a vicious change in direction that would alter our course? But aside from the fact that I had never seen such a quick and total disappearing act of normal visibility, I wondered whether the change in atmospheric conditions did not signal a change in the nature of the weather itself. The fact that it had rained during the night had already raised some doubts in my mind. It was not normal to have rain in the midst of an anticyclone system. Yet fine weather had established itself so firmly with the arrival of day that all my doubts had been dispelled.

Now those doubts returned. That local disturbance could mean a return of easterly winds, the most immediate result of which would be to carry us past the island or, at best, to dash us against the rocky point. What should we do? We were in no condition to react or even to follow the situation as it evolved. We continued to make way, but toward what? There was no way of knowing whether, after having crossed that curtain of mist, we would emerge on another stretch of open sea, on that so much hoped-for beach, or on a wild and unapproachable coast. All that we could hope for was that, as we advanced, we would be able little by little to see more until finally we could see the island itself.

"How did Catherine react in the face of that new situation?"

Catherine: I didn't have any reaction. Lucien's doubts annoyed me. I didn't want to hear about them. What did it matter that the island had disappeared in the fog as long as the wind continued to push us toward it? The

Fifth Conversation: Fog 159

sun made me warm. I refused to have my fragile well-being shattered.

Lucien: To make myself feel better, I reasoned that, given the steady strength of the wind, the fog might well clear up as quickly as it had appeared. There was no doubt that we were covering ground. It was almost noon. Land couldn't be much more than six miles away. It was time to pump again. What an ordeal! It took me more than a quarter of an hour to recover, afterward. There was no point in asking Catherine to help. I saw how helpless she was, reduced to the simple preoccupation of sparing her strength! It was better to let her rest. If we were going to land in a few hours—which continued to remain a possibility—there were a few things she would have to be able to do, especially as I began to have doubts about my own real capacity. That morning's fine optimism had crumbled with the disappearance of the island.

Toward 3:00 P.M., there was further cause for concern. The sky was clouding up in the west. The island-born fog was spreading slowly. Several times during the morning we had seen airplanes heading for the island. We had lost sight of them only as they were about to land. Now we heard them but no longer saw them.

And it was then, when I had seriously begun to wonder whether the wind wasn't carrying us out to the open sea again, that I saw a ship about a mile to our left. Now, that ship was on a course parallel to ours and heading in the same direction as we. For those who knew the area, there was only one possible inference: we were heading northwest, toward Barcelona. The wind had therefore backed into the southeast.

I watched Catherine as she drowsed opposite me,

taking advantage of what was no doubt the last of the sun before it too was enveloped by fog. After all, perhaps I was wrong. I had certainly been wrong about the points of the compass before . . .

Catherine: I'd seen that ship too, and I'd realized we might have lost our way, lost the island, lost our last chance, yet I thought: it can't be, the wind may have shifted but not so much as to make us miss our objective. The fog had muddled our bearings but it hadn't done away with the land. Fog wouldn't prevent our reaching land. We had to be patient, wait until the sun had declined and showed itself clearly enough to allow us to judge our heading. Unless, of course, the island reappeared in the meantime. It shouldn't be that far away. Wasn't the main thing for the wind to hold? What could be worse than a dead calm? As long as there was sun, there was hope.

In fact, Lucien's frame of mind couldn't have been so different from mine at that point because we had started to make plans for the future again. Short-range plans at first, our holiday in Majorca, then more distant plans, my farm, his minibus. I was reticent. All I said was, "I'm through with the sea, it's a little too salty for me." He laughed. We still believed in the future. But I preferred not to talk about it, not to tempt fate. I didn't want to talk about the future.

Lucien: Yes, but when about 5:00 P.M. we realized that the wind, far from dying, was freshening and that the fog was spreading and growing thicker, we really began to worry. We kept telling ourselves, it's going to get better, we'll see land any minute. But we said so with less and less conviction. And when seas began to form and we saw the raft bottom beginning to fill with spray

Fifth Conversation: Fog

that had drenched us, our bogus optimism was badly shaken.

Of course, at a pinch we could always tell ourselves that we were moving in the right direction, that the rising wind would get us there faster. But it was becoming harder and harder to let ourselves be lulled by such illusions. The way the wind was blowing, we should have already reached land by then or, at least, have been close enough to see it no matter how thick the fog. Yet all we could see was the growing whiteness of the sea.

The sun and the blue sky had vanished. We were again plunged into grayness and when, thanks to a sudden clearing, we were able to get our bearings, there was no longer any doubt. The sun, already embedded in thick clouds, was descending to the left of our course. We were indeed heading northwest, toward inaccessible Barcelona.

The Balearics were gone, behind us, lost in the past, like Porquerolles, *Njord,* our lives. We might perhaps cling to all that by streaming the sea anchor in the hope of a change in the weather, a turnabout of the wind into the northwest that would again bring us off Pollensa, but such a hope didn't withstand scrutiny. A northwester was a tramontane which would, if it rose (and nothing seemed less likely), almost certainly sweep us into the channel between Majorca and Minorca and then carry us, to be lost for good, in the direction of the Algerian coast.

Besides, streaming the sea anchor in such weather would have meant our taking the seas full force, letting them break on top of us like a reef. The raft couldn't have withstood it as it would have been swamped in a matter of seconds. There was only one thing to do. Let ourselves be carried by the sea.

I mentioned the idea of the sea anchor to Catherine

to verify my theory. She immediately agreed with me that it would be a hopeless maneuver. She said to me, "There's nothing we can do but wait." Wait for what? For a boat to fall in our lap at the last moment, in broad daylight, like those movies in which the hero is always saved and marries the heroine? Spain, as we've said, was too far. There was no chance of reaching it in our condition.

We had seen land one last time. We had believed for several hours that we might reach it, as if someone had opened a door halfway, had allowed us to glimpse a world to which we no longer belonged, from which we had been exiled for reasons unknown to us. Then the door had been closed again and we had again been cast into the midst of our deserted sea. A bit like the story of earthly paradise. It is shown you for a moment, time enough to picture all its charms and suddenly, without understanding very well why, you are driven out.

The situation had become perfectly clear. We were not wanted anymore. We had to leave. Yet we must arrange to vanish quietly, without too much fuss. What was the best way? I was pondering that question as we prepared to spend our eleventh night as shipwreck victims. For me, the problem was no longer how we could survive but how best to die.

Catherine: Despite my great confusion, I continued to harbor an inexplicable hope, but it was so faint, so tenuous that I didn't know how I could still cling to it. I persisted in thinking that something was going to happen, without really quite believing it anymore.

I thought we had no choice. We had to try to last until dawn. Nothing could happen to us until then. Allowing ourselves that postponement was only putting off the inevitable, but I said to myself that tomorrow was the twelfth day and then I fell into unconsciousness. But not for long, alas! We were not going to be granted a night's grace.

11. The Last Postponement

When Catherine regained consciousness, it was about 9:00 P.M. She had been in a coma-like sleep for less than two hours. The night was dark. She could barely make out Lucien sitting opposite her. Where was the full moon that had lit Majorca so well the night before? Nothing pierced the blanket of clouds that had overrun the sky.

The wind was up to force five. Spray was flying, lashing their faces. The raft had resumed its wild, careening rhythm. Abrupt departure as it rose under a sea, then mad descent into the trough. Fortunately, there was no intermediate phase, as there had been in the storm, during which the raft perched precariously a second or two on the crest of the wave. The sea was unpleasant but not alarming. But then, had it been alarming, what could they have done about it?

Was Lucien asleep? Catherine felt him stir now and then. "I'm thirsty," she said, "shall we have a drink?" "Yes, let's drink," he replied. She took the jerry can. It was light. When it was full of water it was too heavy to drink from alone. One of them lifted it while the other put his mouth to the opening so as not to waste a drop.

For eleven days they had been rationing themselves, trying each time to put off as long as possible the moment of drinking. Then they counted the number of swallows—three, four, five, never more than eight. Sometimes Catherine caught Lucien taking one more than had been agreed upon. She had said nothing, feeling that one swallow more or less would make no difference in the end. After that, she retaliated by taking her allotted swallows in bigger gulps. All that seemed silly to her now. What had been the point of trying so hard to survive if it had come only to this?

She took six big swallows, as big as her mouth could hold and held out the jerry can to Lucien who took it, then, after having helped himself, put the can in the middle of the raft within reach of them both. Catherine understood the meaning of that gesture. They had done the same thing forty-eight hours ago, during the night before their sighting of Majorca. They had been desperate by then. There had scarcely been any reason at the time. In short, they had put the jerry can between them and had drunk immoderately, one after the other. Yet then they had had almost fifteen liters, whereas today there was not enough, no matter what they did, to see them through the following evening.

Eventually, rationing weighed on them. It was exhausting always to have to curb their desire to drink when it was possible to satisfy it.

"When the water is gone," thought Catherine, "the problem will be settled."

She took the jerry can, swallowed six more times, and passed it to Lucien who drank in turn. They had not

The last swallow

spoken a word. Until 11:00 P.M., they sat in silence facing each other, hardly able to see each other in the dark. From time to time, one of them would take the can, drink, and hand it to the other. They no longer counted swallows. They took as much as they wanted. Catherine had the last drop.

"There's a good job done," said Lucien. "Now we can't have any more illusions. It's tomorrow or never. Anyway, I don't think we could possibly have lasted much more than twenty-four hours. Tomorrow morning, tomorrow noon at the latest, we would have run out of water. I'm just as glad it happened now. Things are more definite that way."

"I agree," replied Catherine. Then she added, "You know, I never told you but for a long time I've had the same dream. According to this dream we ought to be saved on the twelfth day."

She had almost said, "I ought to be saved," but had caught herself at the last moment. That was the reason she had never spoken to Lucien of her dream. It only concerned her.

"Tomorrow is the twelfth day, isn't it?" asked Lucien.

"Yes, it's tomorrow."

"We'll see. If you should ever mention your dream to someone else in the future, it would be because it came true. Anyway, we'll know tomorrow. But why tomorrow when we've been waiting so long? You know, I also dreamt often that we were saved, saved by helicopters. It always happened the same way. But I never thought that, because of my dream, helicopters would suddenly appear and rescue us. You always dream about the things you want. That doesn't mean anything."

Catherine insisted. If she had spoken about her dream

it was so that Lucien would help preserve her pitiable hope. She asked, "Did you ever dream we were saved on a certain date?" "No, just that we were saved, that's all."

How difficult it was to cling to something so fragile, to admit that one's life, one's own life, depended entirely on a dream's coming true and continuing to believe in its coming true! Until the tenth day, that business of the twelfth day was a convenient way out. Catherine could always tell herself: if nothing turns up before then, there's always the twelfth day; a sort of double-barreled system that afforded a semblance of security. But now, a few hours before it was there, at the moment when, inexorably, the twelfth day was going to unfold . . . Panic seized her. If only she could stop time, keep the twelfth day from coming.

They were thirsty. Yet, only a short while ago, they had finished their reserve. Lucien pushed the jerry can away with his foot, pumped up the rubber ring, emptied part of the water they had shipped—only part because the pump was almost useless—and then sat down again hoping to sleep. After a while he succeeded.

Catherine was dozing. She was again filled with dreams of eating human flesh. Soon the dreams were as obsessive and precise as they had been throughout Friday and Saturday. Lucien was dead, she was carving him, eating his flesh. It tasted good. Her anguish had vanished. Her survival was assured again. The coast of Spain was within her reach. The twelfth day prolonged itself, endlessly. It was guiding her, protecting her. She knew then that it would never leave her until she was safe, that it would last as long as she needed it. She carved another piece from the body, chewed it. The fresh meat quenched her thirst.

Then something came to disturb Catherine's bliss. Fish were attacking the raft, sharks she thought. Several times they bumped against the inflatable rings, leapt, then nibbled the bottom, no doubt the remnants of the anti-drift pockets that Lucien had cut away.

She heard the sound of teeth against the rubber fabric, felt rough backs rubbing against the canvas. Another blow against the rubber rings directly behind her made her start. She called to Lucien, "Do you hear?" Only half awake, he grumbled, "What's the matter?"

"Fish. Sharks, probably. They're biting the bottom, liable to tear it."

He listened a moment. Nothing happened. "You must have been dreaming. Try to sleep."

How could she follow his advice? The sharks were still there, getting bolder and bolder. To frighten them, to prevent their continuing to bite the raft, she pounded her fists against the raft bottom. The fish scurried off after asserting their reality with a flip of their tails.

A few moments later, they reappeared. She heard the flop, flop noise of their fins striking the canvas. Never had she had to contend with such a strange watery commotion. One of the fish made the raft lurch with a flip of its tail. It was like the sudden movements caused by the waves, long ago, during the storm, only more brutal. Catherine felt the punch of an animal's muscles and again she beat the bottom with her fists.

Lucien, unable to sleep now, heard her fists but no longer had the strength to interfere. He was battling his own obsessions: death, how to achieve death without too much trouble. In a few hours, in the morning, when this endless night would be over, the time would come to put a stop to this voyage, to hope. The simplest way would be to throw himself overboard, the way one leaps into space. The raft would go away, driven by the wind. It would be impossible to catch it, impossible to make it come back. It would be enough, then, to let himself go, to give himself up to the hungry sea. It would be better, though, to remove the life jacket before going over. Everything would go quicker that way.

And so he saw himself, his mouth already gasping to draw one last gulp of air, water filling his lungs, a final spasm, like a disconnected puppet, trying to escape this self-inflicted catastrophe, and then nothing more, a pool of blue in which he was lost, set free from life. Peace at last.

He remembered scenes from a film. An undersea explorer who, dragged down by the weight of his equipment, sank slowly for a long time toward the depths, like a dead leaf falling through the air. How beautiful it was, that slow descent into the abyss. What finer dance could one imagine? The drowning face smiled, lit perhaps by the intoxication of the deep. The limbs had become like seaweed, undulating at the water's whim. Yet it was a woman, not a man. A woman lost in the strangeness, her hair drifted behind her. Who else but a woman could have such elegant gestures, such a way of reclining on that bed of coral.

Catherine was striking the bottom of the raft, still hounding imaginary foes. Lucien stiffened. There could be no enemies any more than friends for the sea was empty. Nothing could touch a body of such icy indifference. There was nothing one could do in the face of such destructive fury. Why not do it now? No, it was night. He ought to wait for day. Something might happen. Besides, how could he face that last moment, that last

gasp? The image of the drowned woman finally recalled him from his resignation.

Catherine was still warring against the sharks that continued to threaten the existence of her last refuge. If they succeeded, then it was all over. She was not surprised to be leading that final fight alone. Lucien no longer existed.

Overcome by fatigue, she drifted from sleep into wakefulness. The swarming of vigorous bodies round the raft continued. She wanted to lean out to try to see them, but the simple business of turning her head would have required an impossible effort. Her torso and limbs were nothing more than an immovable block. Cramps now affected her stomach muscles, adding a general uneasiness to her suffering. At any moment paralysis would reach her diaphragm, her heart. Perhaps that was how it would all end. In a way, it was reassuring. One was always tempted to wonder in what way death would do its work. That way was quite acceptable.

"And yet, in spite of everything," Catherine realized with surprise, "I'm still myself. Exhausted, pushed to the very limits that separate life from death, I'm still here, me, Catherine, with my recollections, my desires, even my plans. Nothing has done me in. No ordeal has put me under. I'm following my path despite all obstacles. Must I not be saved to have the chance to continue? But why me? Am I so essential? Yes. No one else will ever do what I have to do. No one can take my place, ever. I must live. If I perished at sea, without leaving a trace behind me, something unalterable would have happened. I'm still alive. I don't reject anything that may happen to me."

The sharks were attacking the raft doggedly. Their assaults made Catherine angry. "What do they want of me? What interest do they have in my death?" She banged the raft bottom with her feet and fists.

An ugly day dawned slowly, as if with regret, on a gray sea. The wind, which had let up a bit at the end of the night, was reaching force six. The raft was shipping seas. Lucien tried to bail, but without success. The pump was now only effective in reinflating the raft.

No more water, no more pump, exhaustion, bad weather, it really was the end. The only consolation for Catherine was that the sharks had disappeared with the coming of day.

They still sat facing each other, but no longer spoke. They no longer had either the strength or the desire. Catherine vacillated between surrender and resistance, resignation and fear. When it was fear that carried her, all her muscles, already tight, tightened some more. She felt an oppressive ball of anguish in her chest. Every now and then she tried to relax, to breathe normally again, but without success.

But Catherine hadn't lost all hope. The passion for life still burned within her. She still did not want to die, not she, not today, not at the age of nineteen. It was unthinkable. No, she would not die, she knew it. She kept repeating that to herself to convince herself. Something would happen. She would be saved.

Her belief that Lucien would die before her was still strong. And it was only natural. He thought only of death whereas she was bound to life. As long as he was there she could rest easy. On the other hand, if he died, she could eat. She still struggled inwardly with that contradiction. Was it better that he live or that he die? And what if, as her dream seemed to indicate, she could only be saved after his death?

During this time, Lucien had been concentrating all

his attention on the knife. He had made up his mind to end everything. No purpose could be served by prolonging such slow agony. He could use the knife to open his veins, to let life run out slowly . . . that was a sweet death. He pressed the blade against his wrist. Catherine was looking at him. Was he going to commit suicide? She asked him the question. "Yes, I think there's nothing else for me to do."

He's mad, he hasn't the right. Even if one has only one slim chance left to survive, he must take it, wait for it. She said that to him fully aware of the vanity of her words. What hope was there? That a ship would see them? No one had found them during twelve days, except a freighter that had seen their flare. It had left without giving the alert. Had they been told then that they would have twelve days of agony, and that their long wait, their constant struggle would lead to nothing . . .

She looked at Lucien, saw his eyes already fixed in the distance, the tanned skin of his hollowed cheeks, streaked with salt, that stuck to his cheekbones, his unkempt beard, the locks of hair glued to his temples by spray, and the ineffectual penknife in his hand. Faced with the act, he hesitated. He, too, was afraid. "Do you think it takes long? Do you think it hurts?"

What could she say? She had never contemplated suicide. She had never thought about it. After all, it was his business. If he wanted to die, let him take the risks. Yet, to be left alone, even with the food and drink supplied by his body and blood. How would she endure the silence, confronted by her pathetic spoils? She asked herself, "Would I even manage to undress him to get at the flesh?" No, he must live. She needed his presence to survive.

"Are you going to commit suicide after me?" asked Lucien. "If everything went well, could you also do it?"

"No, never! I want to live until the last moment of my life. I don't want to choose what that moment will be. I want to be able to dream of the future until the last moment. Besides, today is my day, the twelfth day."

He still hesitated and she already knew that he had abandoned the idea.

"After all, I can't leave you alone. I must stay with you until the last moment," he said.

That phrase, "stay with you until the last moment," repelled her. It could only mean that he preferred to await her death so that he could eat her, devour her liver, and survive.

Lucien kept the knife handy. His head was wobbling. He told himself it was time to pump, yet between saying it and making the effort to do so . . . No, it wasn't time to pump, it was time to die. He reproached himself for lacking the courage to sever the thread of his life. He put off until later not the decision, which seemed to him already to have been made, but the moment of its execution. In the end, he thought, things would happen by themselves.

What if Catherine were to die first? He thought of his earlier obsession as he looked at her. She was so wretched, crumpled there in the bottom of the raft, scarcely able to keep her body from toppling under the impact of the sea. No, he couldn't bring himself to do it, it was too late, he no longer had the strength to hope, he was too tired to be able to assure his existence at that price.

Catherine was drifting toward similar views. Whenever she succumbed to her fantasies, she was still fighting to survive. During those times, Lucien seemed to her nothing more than a means of nourishment or, worse still, an inopportune presence whose disappearance, as foretold, must precede her rescue. And when she opened her eyes and was again confronted by his presence, all her plans crumbled, swept away by the simplest of evidence. There they were, both of them, victims of the same destiny, both reduced to the state of human debris in which there still flickered an occasional moment of semiconsciousness.

They were still able to form a few more or less coherent thoughts, to come to life for a moment under the whip of a sudden rebellion against death or of anger against their impotence. But they never thought that these were anything more than final flickers before ultimate annihilation.

They were thirsty. They had not drunk for twelve hours now. For the first time their mouths were coated. They could no longer salivate. The wind had freshened. The sea had become really rough. When they lapsed into sleep they were soon awakened by spray hitting them in the face until finally even those assaults failed to provoke the slightest reaction. "They're watching us," thought Catherine, "they're watching us. They're out there, crouching over us. They're waiting for the moment when we're going to die." She didn't open her eyes for fear of seeing the face of her tormentors. "Let them do what they want. I don't have the strength to stop them anymore."

About 2:00 P.M., Lucien woke up. The inflatable arch and the upper rubber ring were almost completely deflated and there was a great deal of water in the bottom of the raft. Without thinking, he took the pump and began to pump up. When it came to the water, it was more difficult. He began by scooping some out with his hands, then tried to use the pump. The pump offered resistance. He became angry and bore down furiously on the pump handle. Suddenly he wanted to overcome that resistance. Centimeter by centimeter, he managed to pump out several liters, then, exhausted, sickened, convinced of the futility of his efforts, he sat down and fell asleep again. Catherine had not stirred.

She woke two hours later. She shook Lucien and shouted, "A ship, a ship, coming right at us." He opened his eyes, heard the noise of the engines, and straightened. The bow of an enormous freighter was fifty meters to windward of them. A moment later it was even with them, passing a few meters away.

They shouted. They saw an officer on the bridge who was looking out to sea, in their direction. Abruptly, he opened a hatch and disappeared. "He's seen us," Catherine exclaimed, "he's seen us, this time I'm sure." Lucien was sceptical. His capacity for enthusiasm had disappeared. For once, he was the one who did not believe.

The ship had already passed them, was far away. But they continued to wave their arms. They could read *Abel Tasman* across the stern. The last chance, it was the last chance. Was it lost like the others? Minutes ticked by. Had the ship slowed down? They thought it had but, from where they were, it was so difficult to judge its speed.

And suddenly, there was no doubt. The *Abel Tasman* made a half turn. It was coming back slowly, gently, making a wide circle around them, as if it were the simplest thing in the world to establish contact with two shipwrecked souls who, a moment before, had already abandoned life, abandoned any hope of ever returning to the world.

Sailors were waving to them from the deck. Lucien burst into sobs. Catherine refused to let herself go, biting her lip. Restraining her immense joy, she added, "It's not over yet. It won't be easy. We won't be safe until we're on board." Soon, the *Abel Tasman* had stopped less than thirty meters away.

Finally, she said, "You see, it's the twelfth day."

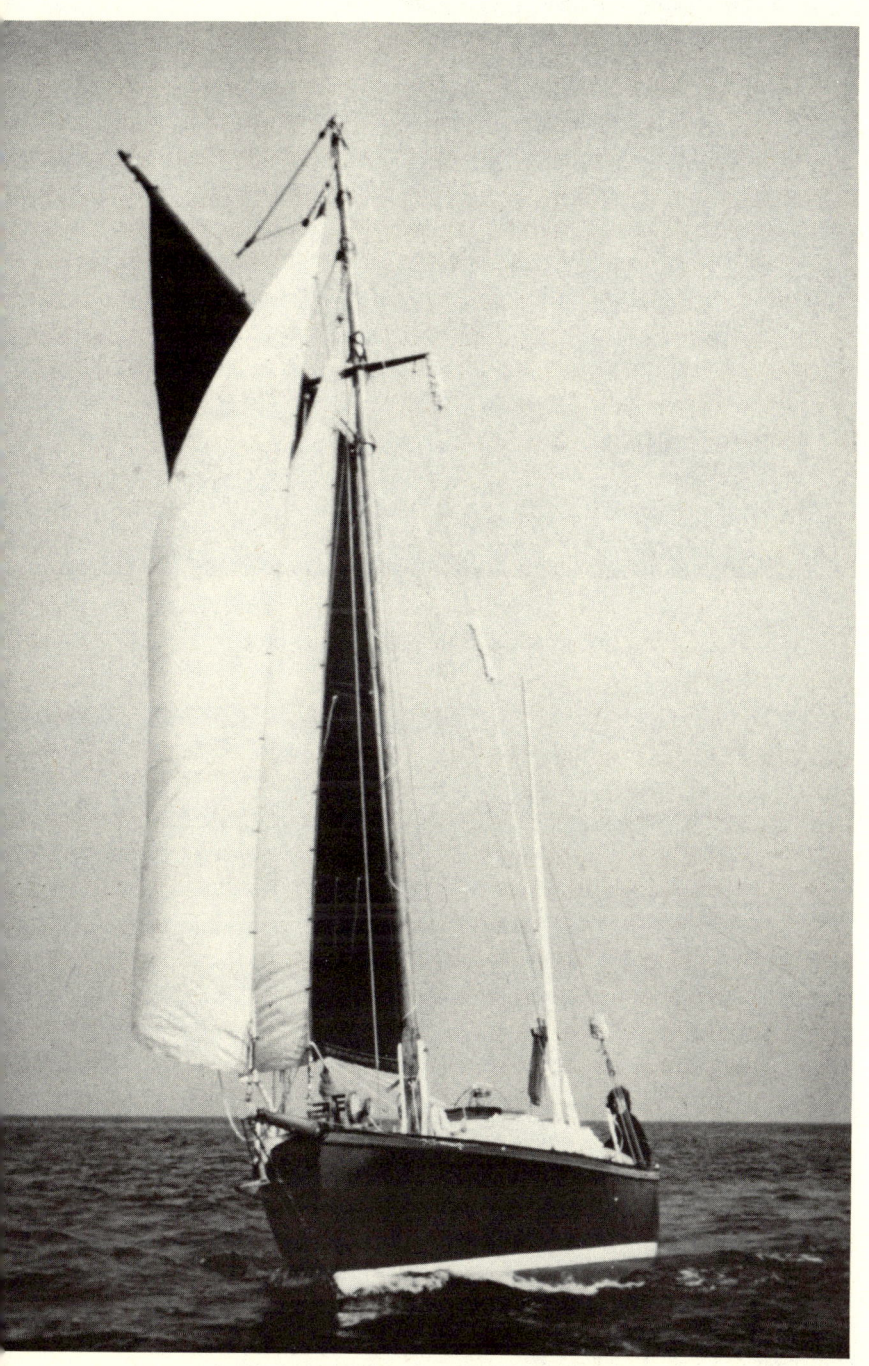
Njord at the time of departing Porquerolles.

Lucien and Catherine: This is the life!

The inflatable raft seen from the *Abel Tasman*.

Saved! For Catherine, a few more steps. Lucien waits in the raft with the sailor who went to help them (Photo: Benelux Press—Richard F. Kaan).

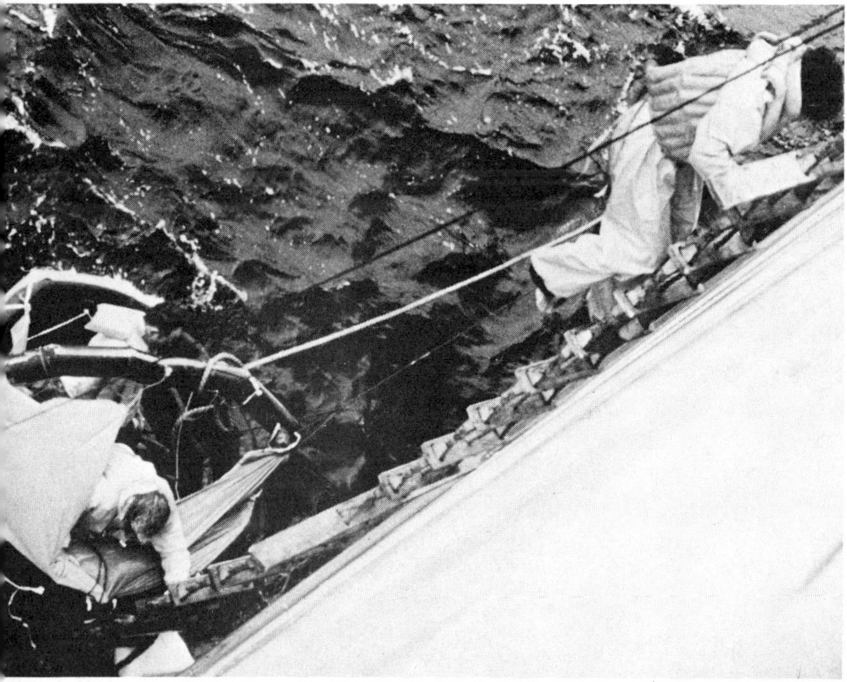

The *Abel Tasman*.

Catherine had only one fear left: that she would fall on the stairs.

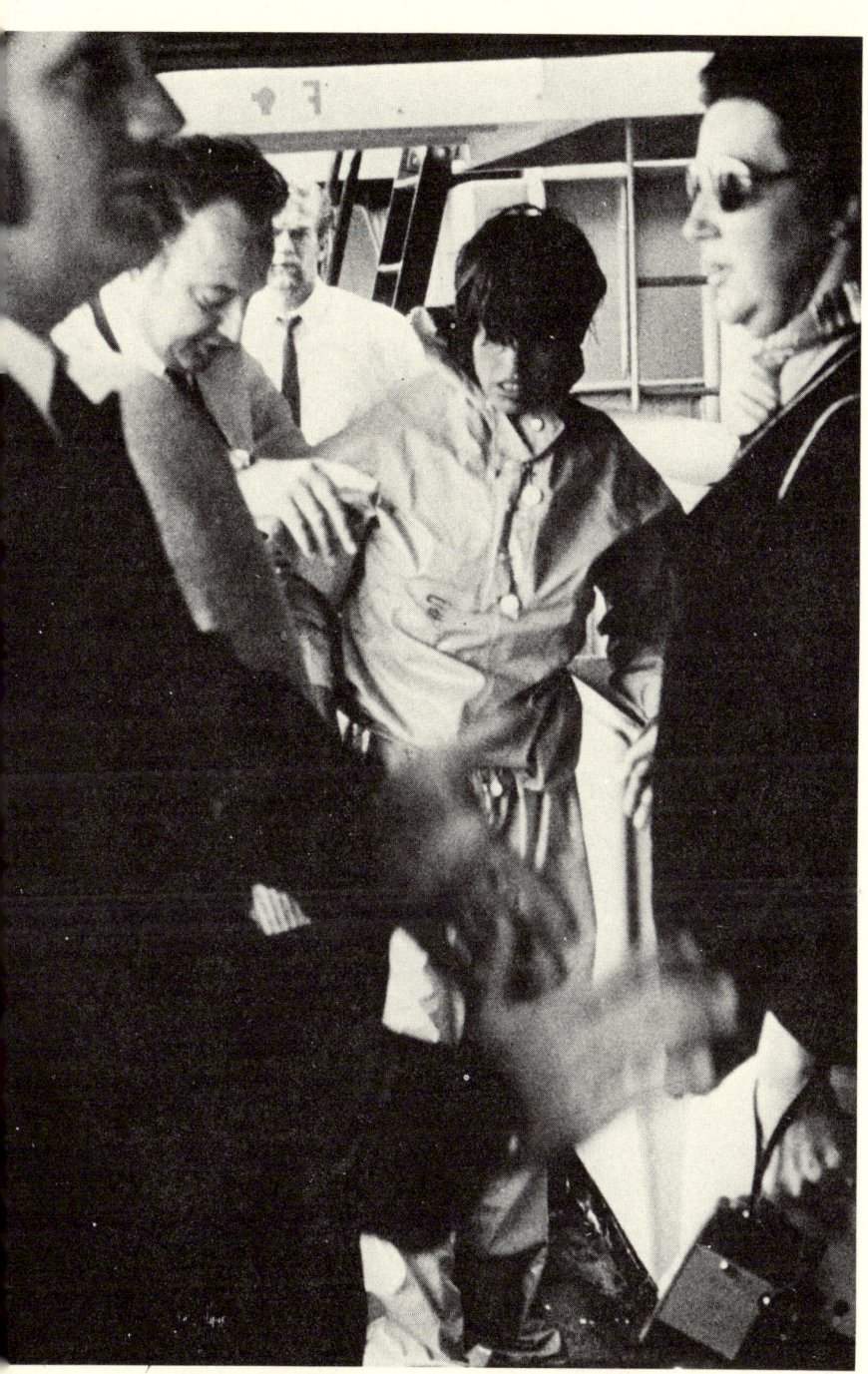

Someone supported me. Who? I have no idea.

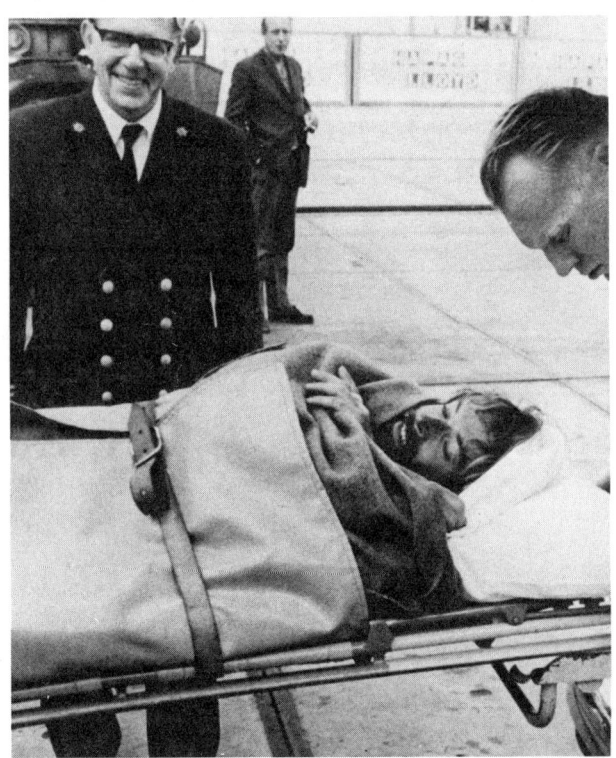

Joy returning, despite the stretcher and bed.

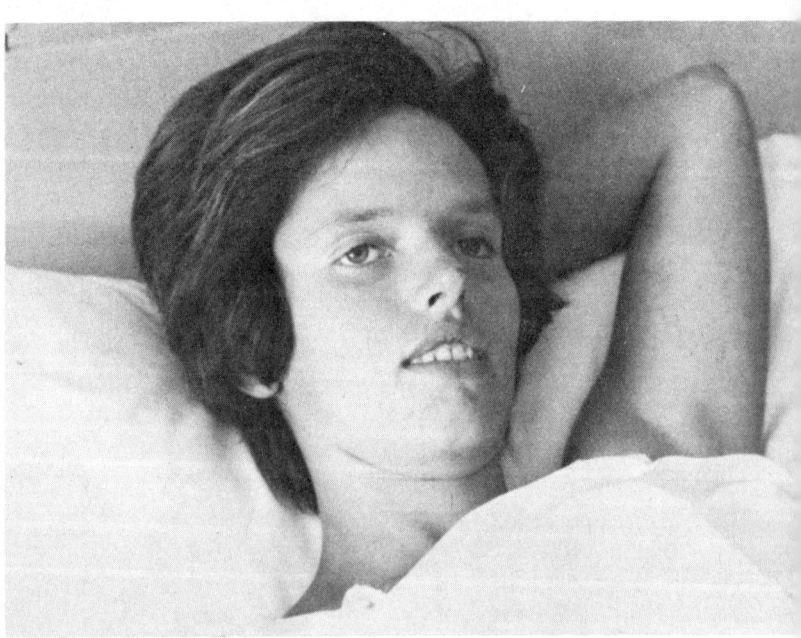

(Photo: Benelux Press—Richard F. Kaan)

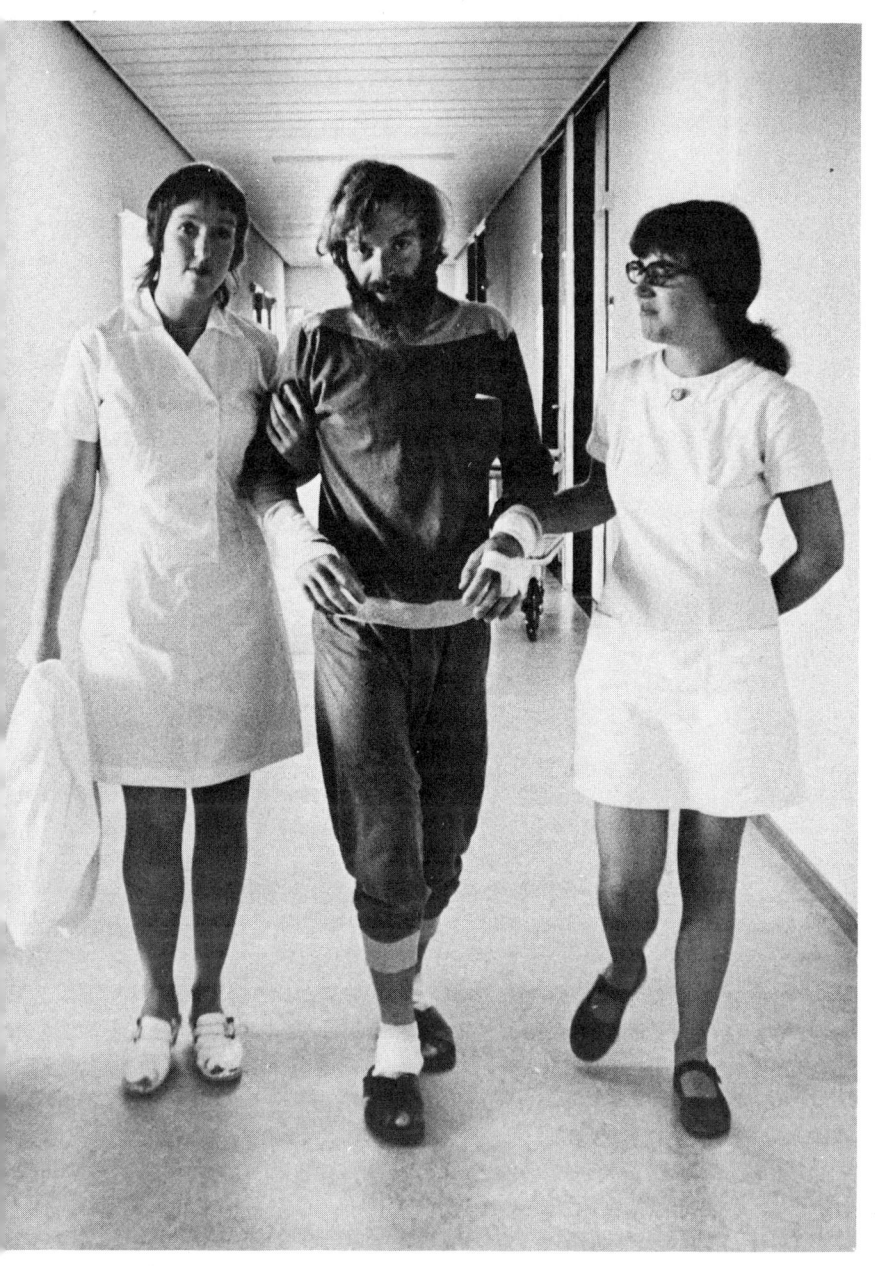
Lucien's first steps (Photo: Benelux Press—Richard F. Kaan).

12. The *Abel Tasman*

"Let's go back a bit to the way your meeting with the *Abel Tasman* came about."

Catherine: We were both asleep. Suddenly I woke up and sat straight up. At the moment I couldn't have said why. No doubt I'd heard a noise, yet I hadn't done so consciously. When I saw the ship, it was about 500 meters from us. It was approaching at great speed yet I was never afraid of a collision. It was obvious that it was going to pass close alongside us. We learned later that it was flying along at twenty-three knots. So it would cover those 500 meters in about thirty-five seconds. In other words, everything happened very quickly.

It was a big ship, 220 meters long, with a completely unencumbered, flush deck for three-quarters of its length and all its superstructure concentrated astern. The sides were a dozen meters high, the bridge stood thirty meters above the water. A huge and impressive thing.

While it was maneuvering to pick us up, we began to ready ourselves for the rescue operation. I was a little worried despite my joy. I'd heard about a pleasure sailor who had drowned while a freighter was trying to recover him. That's why I said to Lucien, "We won't really be safe until we're on board." It was certainly going to be a delicate maneuver, especially given the strength of the wind and the state of the sea. There were troughs nearly two meters deep. The crests were white but the seas were not really breaking. And how was I going to manage with my legs half paralysed? I was kneeling, not to lose sight of the ship, but I felt I wouldn't be able to stand.

Lucien: I told her, "You must move about to try to loosen your muscles." During that time, I began to free everything on board so that nothing would hamper us when the time came. I lowered the sail, or at least what passed for one, I pumped up the arch so it would be good and firm in case we had to lean against it, and I streamed the sea anchor to prevent our drifting. Because it was in our way, I made the jerry can fast to the raft and put it over the side. I'd wanted to throw it away but, I don't know why, I didn't dare do it. I also tried to pump out the water but without success. The pump no longer worked.

After the *Abel Tasman* had stopped about thirty meters to windward of us, an officer, within earshot, spoke to us in Dutch to ask if we could come closer to the boat by paddling. I replied in German that I would try. Thereafter, they always spoke to us in German.

We then set about trying to make the raft go forward but, against the wind and with the little strength we had left, it was impossible. Besides, the presence of that enormous hull caused cross seas that tossed us about terribly. They realized immediately that we would never reach them and they shouted to us right away to stop trying, that they would maneuver instead.

Catherine: It was then that Lucien asked them whether they didn't want us to swim over to them. Fortunately, they absolutely forbade it.

Lucien: Yes, I must admit that the chances of our carrying it off were very slim. But let's also admit that the prospect of colliding with some 10,000 tons was also rather alarming.

When I saw the *Abel Tasman* move full astern and then move forward toward us, I was not happy about it. It seemed to me we stood a good chance of being crushed by the stern, which might well have happened had not the maneuver been carried out so masterfully. We can never say how much admiration we have for the captain of the *Abel Tasman*. That threatening stern came to a stop within inches of us, perfectly upwind, with the gentleness of a ferry making a landing at a quay to let its passengers alight on a beautiful day.

At the same time, two nylon hawsers with knots at their ends came down smartly toward us. All we had to do was grab them, and no sooner did we have them in our hands than the two sailors who had paid them out to us began to haul us toward a ladder that had already been lowered amidships. We held the hawser in one hand and with the other fended off the hull to keep the raft from chafing too badly. For despite the fact that we were sheltered from wind and sea by the imposing barrier of the ship, we were still tossed about a good deal. Our raft rose and fell about a meter against the steel side of the ship. Not only was it rather frightening but it led to difficulties. We were not too steady on our feet, nor were our movements too reliable. We could have lost our grip on the hawser, been pitched into the sea, or had our skulls smashed against the steel hull, and the canvas of the raft ran the risk of being ripped. In short, we had the feeling that nothing was certain yet. We were on the right path, but we hadn't reached our goal.

Catherine: We were so absorbed by the rescue operation that we didn't realize at first that there were twenty people or so on the deck above us leaning over the rail, looking at us, and taking snapshots. There were both men and women. I was surprised. I didn't think there would be women on a freighter like that. I learned later that they were officers' wives.

Once having come abreast the ladder, we clung to it, yet it was very hard to hold on to it because the raft moved so much. We were thrown about every which way. Lucien wanted us to climb right away, but they shouted to us to do nothing. A sailor climbed down. He struck me as being enormous, a real colossus. Actually, he wasn't exceptionally big. He was the ship's carpenter. When I saw him again afterward, I didn't recognize him. I was surprised to see he was of normal stature.

He made the two hawsers fast to the lifeline that ran inside the raft while we continued to grip the ladder. I must confess that from then on my recollections are muddled. I know that I was strapped into a sort of harness with a hook at the back and that I was hooked to a cable that had been lowered by a winch. Then, it was explained to me that I had to hold on to the ladder while I was being lifted, as if I were climbing it myself. And immediately afterward, I felt I was beginning to rise. It was a curious feeling. The rungs filed past me, I worked my arms and legs but without having to make any effort and yet I rose. Once level with the deck railing, I was seized by another Hercules—but this time he really was enormous, I was able to verify it later—who set me down on the deck.

I was being supported. By whom, I had no idea. It

was a good thing, though, because my legs couldn't hold me. I might have had the strength, but my sense of balance had totally deserted me. The deck was firm, I could press against it without its sinking. For such a long time—twelve days is a long time—I'd formed the habit of walking, or rather of dragging myself, across a flabby surface that yielded under my weight. And then, the freighter was motionless. Perhaps it rolled slightly to the swell, but I was unable to feel such faint movements. The world I had just left, whether the raft or even *Njord*, knew nothing of stability. That world, even during periods of the greatest calm, was animated by a perpetual rocking. They say that when sailors go ashore after a long time at sea they sometimes are land-sick. Well, I had it, to the point of vomiting. But perhaps the greatest shock I had was feeling the vibrations of engines despite the impression of motionlessness conveyed by that floating rock, the *Abel Tasman*. The vibrations rose all the way through me from the steel deck. It was like a physical presence, the presence of life itself.

At the same time, there were all those people crowding round me, congratulating me, asking me questions. I felt arms holding me up, strong, healthy arms, that had not been slowly gnawed by seawater, that had not been reduced to fragile tentacles by hunger. I was thanking people, I was thanking everyone, I was stammering, I think. Sentences tumbled out without any order, in French, in English. I was mixing words, languages. I had been saved. I was deeply moved. I tried to hide my emotion under a flood of incoherent words. I wanted to hug everyone. My head was spinning a little.

How long was I like that? Not long, scarcely a minute, I should think. Someone came and lifted me onto his shoulder, my head forward, my legs to the rear. Another colossus. That boat had a crew of giants. Carried in this manner, I crossed the deck, went along passageways, climbed ladders. I saw the deck pass under me. I thought, "If only he doesn't let me go, if only he doesn't fall." That was my ultimate fear—smashing my face on a flight of stairs. He put me down in the infirmary. There were four women there. They undressed me and put me in what seemed a boiling hot bath. I was in their hands, there was nothing further for me to do but abandon myself to their care. Then I looked at my body and it frightened me. Next to those healthy beings I suddenly realized that I had become a dying skeleton, covered with wounds and whose skin, where it was intact, was a strange reddish color. I didn't have the courage to ask for a mirror to look at my face.

Lucien: During that time I also was living through the experience of being rescued. I had seen Catherine disappear behind the guardrail of the *Abel Tasman*. I knew that I wouldn't have to worry about her from then on. But, all those who had been watching us disappeared at the same time as she. From then on the show was not, at least for the moment, in the sea. Well, I'd had enough of clinging to that ladder in the raft that kept rising and falling against the hull. I said to the sailor, "I'm going to climb up, I don't need anyone," and, before he had time to make a move, I started up the ladder. Oh! Not for long. My legs shot out from under me and he caught me just in time. Without him, I would have been in the water.

He then helped me put on a harness. The cable from the winch was lowered again. I was hoisted through the air along the ladder. I was lifted clear of the deck railing, I was on the deck and then I fainted. When I came to, I was in the infirmary, sitting in an armchair. I was given a stiff whiskey. I closed my eyes and took a sip.

"A whiskey!"

Catherine: Yes. I was given one too, but I asked for a very light one with a lot of water. I drank it in my bath. I felt well, a little dazed, but well. I had the impression my muscles were unknotting at last. I heard people asking me what exactly had happened to us. When I said, twelve days, I realized they didn't believe it. They thought I was delirious. They admitted it later. Twelve days? Impossible. Three or four, perhaps. It's true that I must not have been expressing myself very clearly. They never dared say that I was raving, that all I did was repeat the same thing over and over, but I know that's what they thought.

When they carried me back to the room, after the bath, I found Lucien drinking his whiskey. He looked at me and said, "You know, it's good." He was sitting, fully clothed, still wearing his oilskins, and with a blanket over his shoulders. At the time, it didn't strike me that way, but later, when I thought of him in that condition —poor, disintegrating wreck with his whiskey glass and a blanket over his oilskins—I couldn't keep from smiling.

At the time, I was in no state to cultivate a sense of humor. They made me lie down and began to coat my body with several different ointments while Lucien was being undressed. I'd seen him naked less than forty hours ago, when we had bathed within sight of Majorca on Sunday. Events had piled up so quickly since then that it seemed as though time had counted double or triple. I had the impression that more time had elapsed in the two days between our approach of Majorca and our rescue by the *Abel Tasman* than between the end of the storm and our sighting of land. Well, in short, I realized then that his condition was far more serious than mine. His arms in particular were completely raw. They had been in direct contact with wool, for he had worn a sleeveless undershirt. Let's not talk about how thin he was. I thought, "He looks like an old fakir."

Lucien: It wasn't just my arms, either. Seawater had dug into the burns on my hands, which had been caused when I had ignited the flares and sticks of phosphorous. I had deep cuts on my legs, buttocks, and on my heels. Yet the bath did me good, too. The alcohol that I'd taken was making my head spin. I wasn't too sure where I was, what I was doing, or even who I was.

Catherine: Oh! Speaking of who I was, a strange thing happened then. They came to ask me my name and address so they could let my family know. I was adrift in a thick fog. When they asked me that question, I thought, "I don't want them to get in touch, everyone will be alarmed." It was obviously stupid. Yet suddenly, I was unable to answer, as though I didn't know who I was anymore. So they asked Lucien for my name and address, but he must not have been too alert either because the information he gave them about me proved to be worthless. As a result, his family was informed forty-eight hours before mine that we had been saved. Now, what I didn't know was that a week earlier, on the 19th of September, *Njord,* still afloat, had been spotted by an English pleasure sailor who had given the alert. Then, on the 28th, two days after we had been picked up, Lucien's parents and mine were told, as a result of the inquiry into what had happened to *Njord,* that we had perished. So Lucien's parents learned that he had been saved before knowing that his boat had been found empty, whereas my parents were informed that we had disappeared two hours before hearing that I had been found safe and sound. In a case like that, two hours can be a long time.

Well, to continue, I was in my bed and Lucien was in his and, to tell the truth, I think we both felt equally badly. My buttocks were particularly painful and I was unable to move. An officer came to examine me. He put a thermometer into my mouth. Then, when he looked at it, I saw his face go white. I had a temperature of forty-two degrees centigrade.* He telephoned the captain who told him, as I learned later, "There's only one thing to do in a case like that. Use a different thermometer and take her temperature elsewhere." Actually, I had only forty degrees the second time. My blood pressure was very low. I don't know exactly how low it was, but apparently it wasn't good. I felt as if I were boiling, like in a sauna, and I had the impression my blood was beginning to circulate properly again and that my feet were being bitten by ants. Those ants were a tenacious lot, they kept it up for fifteen days.

They told me they would get in touch with a doctor in Holland. They could have done as they pleased, I was in no condition to react. Yet, I was disappointed when they came to tell me that, if the fever did not go down soon, they would have to stop at Gibraltar and put us in a hospital. I felt snug on that ship. I wanted to stay there. Besides, the boat was going to Fleminguan, Holland, next door to Lucien's home and not very far from mine. It was due to arrive there in four days. Four days! Long enough to recover. It was then 5:00 p.m. We had first seen

---

* 107.6° Fahrenheit.–Ed.

the *Abel Tasman* about 2:00 P.M. The rescue operation had lasted nearly two hours. Therefore, we had been on board roughly an hour. We were to reach Gibraltar about 10:00 P.M. I willed myself to get well by then. As it turned out, by 9:00 P.M. my temperature had dropped to near normal. I'd had a heat flash, as they say.

Lucien: I wasn't exactly in the pink, either. Although ointments and dressings eased my suffering a little, I was in a good deal of pain. But I had no high fever, barely thirty-eight.* On the other hand, my feet were so swollen I had the impression they were going to burst. And for more than a month afterward, I was unable to put on shoes. And to think I once used to sell them!

They brought us water and milk. We drank a lot of it. It was delicious. Then they gave us some soup, two bowls each. After that I fell so deeply asleep, Catherine did also, I think, that when I woke up two hours later I thought it was the following day. From then on, the normal pattern of existence began to reassert itself bit by bit. Eating, napping, sleeping, putting up with discomfort, pain, nightmares, but also rediscovering a thousand pleasures one thought gone for good. The first fruit juice, the first cup of coffee, the smell of a dish, the taste of meat, and at the same time the happiness of being able to make plans without first thinking: if we're still alive . . .

Catherine: Yes, it was that above all. To feel again master of one's fate. I saw people. I began to tell them my story, fully aware that I was still babbling disjointedly. My head was full of thoughts, memories, plans. I could scarcely walk but my spirit was leading the fight. I had

---

* 100.4° Fahrenheit.—Ed.

learned that at the time of our rescue, on Tuesday, we had been seen not only by the officer on the bridge but also by all those who had been in the dining room and by two sailors who had been working forward. A regular festival! When I thought of all the ships and boats that had ignored us during the preceding days, it seemed as though something which had hidden us until then had lifted on that Tuesday when the *Abel Tasman* had appeared.

I had also learned that the wind had freshened again after our rescue. About 9:00 P.M., the ship's anemometer had registered a steady force eight that continued all through the night. How long it lasted no one knew because we left the Mediterranean after that. Although once aboard our floating palace we had not been aware of the wind, we had no trouble imagining what would have happened if we'd had to contend, weak as we were, with another bout of bad weather on board the raft.

I was getting better. Thursday, two days after our rescue, I had the strength to explore the ship, though I hadn't yet fully regained my sense of balance. It made me think of the legend of the little mermaid, the one who was transformed into a woman and couldn't manage to walk once ashore.

By contrast, Lucien remained in very poor condition and was unable to walk without help. At Fleminguan, journalists, photographers, an ambulance, and Lucien's parents were waiting for us. Each of us was going home. I was going straightaway, Lucien after a week in the hospital. The journey was over. We had been given the chance to live.

Commentary

Throughout the reconstruction of the adventures of the *Njord* survivors, I have striven to be both a faithful reporter and an impartial inquirer. What had to be done was to convey the truth about what Catherine and Lucien had gone through between the 15th and 26th of September 1972. My role was that of intermediary, and I adhered to it. At no point did I allow myself to intervene in the unfolding of the story except to put pressure on my "characters" in order to force them to reveal the entirety of their experience, and then to use my own mode of expression to translate that experience.

Obviously, this did not prevent my reflecting upon the events I was reporting and trying to learn something from them—which I'm sure was often the case with the reader, too.

I was inclined to think it was unnecessary to add those thoughts to the account; the story of the shipwrecked survivors of *Njord* was enough in itself. But several friends managed to convince me to the contrary. Thus I add my comments to this story. If I am able to make some contribution toward the problem of survival at sea, I hope my intervention will be forgiven.

Njord

It was a stout boat with a steel hull that Lucien had carefully restored. But she weighed seven tons light and no doubt eight tons fully rigged and fitted out. That was too heavy for a boat eight meters long. Excess weight is the curse of steel hulls under ten meters long. A boat too heavy for its size rides seas poorly. It fights the water instead of going with it. This makes it more vulnerable.

The *Njord* had two weak points. A wooden rudder and a worn storm jib which both gave way. There is no point in having overbuilt equipment if it conceals flaws.

The boat lacked an automatic steering device. Lucien had on board the necessary materials to build a steering vane, but, anxious to put out to sea, he had decided to put off its construction until their stay in the Canaries, when they would have been equipping themselves for the long Atlantic crossing. He had thought that relatively short journeys would allow them to take turns at the tiller.

In theory, he was right, even though the leg from Gibraltar to the Canaries (700 miles) or from Casablanca to the Canaries (540 miles) was quite a distance for a crew of two to cover. As it happened, even a 240-mile crossing (Porquerolles to the Balearics) in bad weather without a steering vane proved that such a distance could severely tax the strength of a two-man crew. By the evening of Friday the 15th, Catherine and Lucien had been at sea for sixty hours. During that time they had hardly slept or eaten and had been badly tossed about. Certainly the extraordinary resistance they displayed thereafter amply proves that they were far from being at the end of their strength at that point. Nevertheless, they recognized that fatigue had clouded their minds somewhat at the time of abandoning ship. The presence of an automatic steering device on a boat manned by fewer than three people during voyages that may last longer than forty-eight hours is an important element of security.

It is a troubling fact that the boat, left to its own devices, survived the storm. Should they have stayed on board? We will deal with that aspect of the problem when we come to the abandonment itself. For the moment, let us content ourselves with what we know. *Njord* was sighted by an English pleasure sailor, Mr. Patrick Chilton, who was sailing aboard the yacht *Adelina*. Mr. Chilton drew alongside the derelict. He climbed aboard *Njord*, searched it, and found the ship's papers glued to the cabin ceiling. The cabin was half full of water, so full in fact that at one point Mr. Chilton thought the boat would go down under him. There was much damage below decks. It wasn't possible to take *Njord* in tow. Mr. Chilton therefore confined himself to noting the position (40° 25′ north by 04° 25′ east) and sending out an alert.

Ten days later, *Njord* was still afloat. She was sighted on September 29th by a Danish freighter, the *Peder Most*. It was then 40° 28′ north by 02° 25′ east, or less than twenty miles from the spot where, three days before, Catherine and Lucien had been rescued by the *Abel Tasman*. Had *Njord* been following the same course, at a slight distance, as the rubber raft? It is impossible to tell because, as we shall see, it is not possible to determine the exact course of the raft.

All we can say for the moment is that, if Lucien and Catherine had stayed aboard and had done nothing to influence the boat's drift, they would in any case have been rescued on the 19th, or the fifth day, by Mr. Chilton. But we are dealing with an academic hypothesis. How can one imagine the ways things would have happened had the crew remained on board? Besides, it should be pointed out that between September 15th and the moment when *Njord* disappeared one way or the other, it was sighted only twice.

Had no one else seen it, even though a sailboat with its standing rigging is easily spotted on the water? Or had some seen it and thought it was just a boat on a pleasure cruise? That it had gone unnoticed by anyone is in any case most unlikely, and it is astonishing that a boat bobbing under bare poles and abnormally low in the water should have aroused so little curiosity.

The Abandonment

Here we are broaching a delicate matter. Could Lucien and Catherine have prevented their misfortune? Can one learn something from their behavior on board ship? It is very easy to criticize from one's armchair after the fact, to say that—as the events which followed seem to confirm—under no circumstances should they have abandoned their boat until it went down. Yet, at the time, faced with the actual situation, things were quite different. In view of the fact that another breaking sea could well have sunk the disabled *Njord*, Lucien and Catherine were certainly not prompted by panic. It was hardly possible not to take into account such a contingency nor even to dismiss its great likelihood. At that point, there was nothing left to do but to prepare for disaster and especially to ensure that, should they need it, the inflatable raft worked, which was apparently only possible by actually inflating it.

Here, it seems, is where the fault lay. A knowledge of this adventure provokes an initial reaction. If the inflatable raft had been tested before leaving, Lucien would not have had to inflate it until it was absolutely necessary. Yet, as one thinks about it, one wonders whether things were not more complicated.

Imagine yourself, at the close of day, on a raging sea aboard a disabled boat likely to sink at any moment. Are you so sure of being able to wait until the last moment when the least failure, human or mechanical, would prove fatal? What would have happened during that last moment? We don't know. A few moments earlier, Lucien and Catherine were hurled into the sea by a breaking wave. It was thanks to their strength alone that they managed to climb back aboard. What would have happened had the boat gone down then?

Besides, it was not enough to inflate the raft. Provisions and necessary equipment had to be put aboard, which took time. The more one thinks about it, the more one comes to the conclusion that, under the circumstances, the course of action followed by Catherine and Lucien was not unreasonable. It would have been sensible had there been some secure way to attach the raft to the boat. But the raft had no provision for making fast a line.

Can one criticize Lucien for having streamed the sea anchor astern when fatigue compelled Lucien and Catherine to heave to in order to be able to snatch some rest in the cabin? Streaming the sea anchor from the bow would have reduced greatly the risk of being swamped by seas. Yet it would not have prevented cross seas from putting the boat on its side, even to the point of laying its mast in the water. Let us not forget that while the sea anchor was streaming astern, the boat took seas only in its cockpit. Had Lucien decided to ship the sea anchor and run before the wind under bare poles, he feared lest *Njord* again be laid on her beam and even be rolled over completely with the risk, among other things, of being dismasted. Lucien ran the same risk had he streamed the sea anchor over the bow. Had he then decided to run before the wind, he was likely to be swamped by breaking seas.

Indeed, the two alternatives available to a sailor in distress, heaving to and running, are both fraught with danger. These maneuvers, practiced one after the other, lead to serious damage and threaten the very survival of the boat. For any given boat and crew, there are times at sea when it becomes impossible to master the situation. Catherine and Lucien were faced with such a moment. They could only hope to survive. What is admirable is that they did survive.

The Journey

What was the exact course of the survivors during the 260 hours they spent adrift? All we know on the subject is the approximate starting point, that is to say the position of *Njord* when it was abandonned (41° 20′ to 30′ north by 5° east), and the exact position of the inflatable raft on September 23rd, within view of Majorca, and on September 26th, at the time of rescue. Aside from that, as we have seen, we have reliable information about the position of *Njord* herself on the 19th and the 29th of September.

Catherine, Lucien, and I debated at great length the question of the raft's course while together we strove to reconstruct the events, a more involved undertaking than one might suppose. The monotony of the days, especially from the end of the storm to the sighting of the Balearics, the great weakness of the shipwreck victims, had created a confused world from which emerged here and there a precise recollection, not, however, always agreed upon at first by both of them. And, even when there was agreement about a fact, there was not necessarily agreement about its chronology.

Now if we were able, little by little, to eliminate the confusion and contradictions concerning the events themselves and to reconstruct the story as it actually happened, we were not able to do the same when it came to the course. Lucien and Catherine said so themselves. During their six days of easterly drift, they had only the vaguest notion of where they were. They even wondered whether they were to the north or the south of the Balearics after having thought for a while they might reach Corsica, which would have put them considerably to the north of those islands. Once they sighted Majorca to their south, their doubts were dispelled and they concluded, after the fact, that they must have drifted more or less along the fortieth parallel. Learning later the two points at which *Njord* had been sighted only confirmed their opinion which, however, gave rise to several objections.

One may ask, for instance, why the raft drifted proportionately so much less rapidly during the storm than subsequently, especially during the last thirty-six hours.

Commentary

It is true that there may be an explanation for this anomaly as, after the storm, Lucien had cut off the antidrift pockets and had rigged a jury sail, which might account for the increase in speed.

According to this hypothesis, nearly 300 miles were covered. During 260 hours, that represents a little better than an average of one knot, which is plausible if one takes into account periods of calm.

Thereupon René Mayençon enters the picture. M. Mayençon, armed with his meteorological arsenal, destroys our fine hypothesis in one fell swoop: "A westerly course along the fortieth parallel, at that time, was impossible. A contrary wind was blowing. To have encountered easterly winds the raft would have had to have been near the thirty-ninth parallel, about sixty miles farther south."

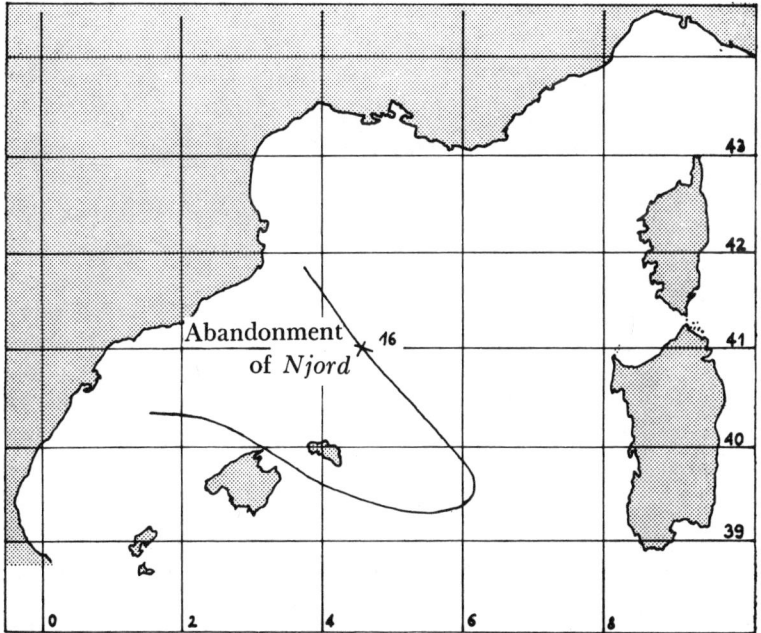

His argument is irrefutable, and the course followed as a result of the winds is in many respects satisfying to the intellect. But if his course is accurate, then the raft must have covered 600 miles in twelve days, or an average of fifty miles a day, which seems a great deal.

Where does the truth lie? We will probably never know. At the very most we can prudently set forth a compromise hypothesis. Perhaps on the evening of the 16th when Lucien and Catherine abandoned *Njord,* her position was farther south and farther west than we had estimated, for example 41° north by 04° 5′ east. When sighted three days later, her position (40° 25′ north by 04° 25′ east) would not contradict that supposition.

After that, the raft, swept along by the storm, would have passed much closer to Minorca than one had thought, and would have continued south until about the thirty-ninth parallel, thus making a journey covering a distance of 110 miles in forty-five hours. Then, blown by easterly breezes, it would have drifted as René Mayençon suggests, but starting from a decidedly more westerly point. That would make a voyage of some 350 miles, which, in terms of distance, is more likely.

The Inflatable Life Raft

"My concern arose," writes Alain Bombard in the preface to this book, "over the failure of the inflatable raft." And he spoke of the need to recall ineffectual and dangerous rafts from circulation.

It is indeed extremely disquieting to know that a reputedly safe survival device should capsize with such unfailing regularity during the storm. The seas were big and dangerously steep, to be sure, but they were not strictly speaking of exceptional size. Storms with winds reaching force ten and waves six meters high are, after all, fairly frequent and, explosions and fires aside, one does not usually abandon ship in fair weather.

The first question that arises then is the following: why all those capsizes? For Alain Bombard, there was only one explanation. The surface area of the raft was too small. But what is the minimum surface area that will allow a life raft to surf no matter what size the wave, to remain stable under any conditions of wind and sea? That problem has not yet really been solved.

Let us simply point out that the diameter of the raft used by Lucien and Catherine was less than two meters. It was an old model made in England, bought second-hand from an English pleasure sailor. Let us also point out that *Njord* flew a Luxembourg flag and was therefore not subject to French regulations governing the safety of pleasure boats.

But while French safety regulations are meticulous, do they achieve their purpose? That is the second problem to be solved. In certain respects, observing those regulations would have had salutary consequences for the crew of *Njord*. For example, it would have obliged Lucien to have checked whether the raft was in working condition and properly equipped *before* leaving. Among other things, the raft should have had sufficiently sturdy mooring points to permit making it fast and towing it. For it was the absence of such points that brought about the raft's parting from *Njord*. But then one may question whether any mooring fittings, no matter how sturdy, would have held up for long under such conditions.

Also, according to French regulations, there must

be a boarding ladder on board. We have seen that the lack of such a ladder often put Catherine and Lucien in a difficult position.

But aside from the fact that the French codes covering "the conditions of approving and checking the life rafts of commercial, fishing and pleasure crafts" contain numerous ambiguities (especially concerning the minimum necessary surface area), one realizes, when one studies that ordinance closely, that having the prescribed life raft for the given size and maritime category of one's boat in no way insures one against the dangers inherent in abandoning ship.

The main criticism to be leveled against that system is that it leads too easily to the impression that having the prescribed equipment is in itself enough to ensure a good chance of surviving. We shall see that it ensures nothing. And that is the fault neither of requirements nor of prohibitions.

Today, any vessel flying the French flag, no matter how competent the captain and the crew and regardless of how the ship is fitted, may not undertake a longer voyage than that for which she has been approved. If you are in the fifth category you are not allowed farther than five miles from shelter. In the fourth, you can go as far as twenty miles; in the third, as far as one hundred miles; in the second, 200 miles. In the first category, you may go where you please.

Similarly, inflatable life rafts have been divided into four classes according to their specifications, the first class being the most exacting. You must have a Class I raft if you are navigating under the provisions of category one. If you are navigating in the second or third category and your boat is longer than ten meters, you must have a Class II life raft; but if your boat is less than ten meters long, a Class IV raft will do. In the fourth and fifth categories, inflatable life rafts are not required. Class III life rafts are not used in pleasure boating.

Thus, depending on the size of your boat and the kind of navigating you are doing, in case of abandonment you would be confronted by different types of life rafts, although for the most part you would find yourself facing identical conditions of survival. Who would dare to say that the problems of survivors lost one hundred, 200, or more miles from shelter are not very much alike? The adventure of Catherine and Lucien bears witness to that fact.

How do these several kinds of rafts differ? First of all, they differ in construction.

In Class I, the tent must be made of several thicknesses of waterproof cloth or have a double wall. It must have a permanently attached receptacle for gathering rainwater. There must be a double bottom to protect flotation and to insulate occupants from the cold. None of that is required in Class II; the tent and bottom can be of a single thickness. As for Class IV, a self-erecting tent is not even required.

It is high time, by the way, to settle the account of that infamous Class IV which is life raft in name only. Its chief characteristic is that it can be put to mixed use. On crossings, it is equipped as a life raft; during stays in port it can serve as a tender. It is a rectangular affair of classic design easily maneuvered by oars and even capable of accepting a small outboard motor on a transom mount. Its principal advantage is that one comes to know it well through regular usage—its construction, its equipment are soon familiar. One knows exactly what one will find aboard in case of an emergency, which is very little, but at least one knows ahead of time. Its chief drawback is that one knows, barring some miracle, one cannot rely on it to save one in a heavy sea. Yet, let us repeat, this very raft is prescribed in the second category if the boat is less than ten meters long.

One can understand the reason that led to the adoption of a compromise solution—the desire to spare the owners of small pleasure craft the encumbrance and financial burden of a tender and a life raft. Yet the chief result has been to encourage owners to buy an uncomfortable tender doubling as a life raft very likely to prove ineffectual. Some claim that the principal advantage of such a dinghy is that it allows one, in the absence of help, to reach land by one's own efforts. But aside from the question of equipment—about which more later—giving a very theoretical aspect to that possibility, we don't see why what is deemed good for the occupants of boats under ten meters long should not be good for those of longer boats and vice versa.

We just now mentioned the problems of equipment. Let us look at them closely. You will find below a list of equipment and provisions which must under law be found in inflatable rafts. One would have little trouble drawing the following conclusions.

In Class I, note the absence of a water pump, indispensable for the rapid emptying of the raft. Also note that the required number of parachute flares is dangerously diminished. To do things properly, there should be four.

The food rations are notoriously inadequate; they would scarcely provide enough for two days.

The water rations are still more inadequate. Even

EQUIPMENT FOR LIFE RAFTS USED IN PLEASURE BOATING

	CLASS I	CLASS II	CLASS IV
Small rescue marker [1]	1	–	–
Light floating line	30 meters	No specified length	
Floating knife	1 or 2[2]	1	1
Flexible bailer	1 or 2[2]	1	1
Sea anchor	2	1	1
Paddles or oars (pairs)	2[3]	1 or 2	1 or 2
Sponges	2	2	2
Repair kit	1	1	1
Air pump	1	1	1
Instruction manual	1	1	1
Fishing kit	1	–	–
Signaling mirror	1	–	–
Whistle	1	–	–
Waterproof flashlight	1	1	1
Extra batteries	1 set	–	–
Extra bulb	1	–	–
Parachute flares	2	2	2
Red hand flares	6	3	3
Table of distress signals	1	–	–
Food rations (per person)	2,250 calories	–	–
Water (per person)	1,500 cm.^{3}	600 cm.^{3}	–
First aid kit	1 for 12 persons	–	–
Antiseasickness pills	6 per person	6 in all	6 in all

[1] We have italicized indispensable items not prescribed for all rafts.
[2] According to the number of persons allowed on board.
[3] If oarlocks are provided.

under a system of stringent rationing, the supply would be exhausted after two or three days. And what assurance is there that rescue would take place within that interval of time? Furthermore, in the event of a downpour, how is one to collect rainwater?

Although a fishing line is prescribed, where is one to procure the bait needed to attract fish?

Class II requirements are woefully inadequate: 600 cubic centimeters of water per person (against one and a half liters in Class I), no food, three hand flares instead of six, and only two parachute flares. And, especially, no signaling mirror, which can prove even more useful than flares and is hardly cumbersome. The fishing line has also disappeared, and let's not overlook the lack of any device for collecting rainwater.

In Class IV, there is the minor omission of any water rations whatsoever. Now we are dealing with the situation that confronted the survivors of *Njord*—the starkest possible conditions of survival. In fact, the "legal" Class IV life raft is less able to cope with seas than that of Catherine and Lucien. It has not been established that one can erect and secure the tent without trouble in a strong wind and heavy seas—and a very primitive tent, at that. Finally, the living space, because of its shape, is severely restricted. To impart motion to the raft, theoretically designed for that possibility, there are two small double-ended paddles far too short to be of any use. It should be possible to rig a centerboard and sail, but there is no provision for either in the standard equipment.

That's the picture. It's easy to conclude that a shipwreck victim having at his disposal a Class I life raft will be far better off than one who has a Class II raft or, still worse, a Class IV. First, he will be less cold, thanks to the double bottom and the greater thickness of the tent material. Then he will have some food and water. He will be able to fish, to collect rainwater and will be aboard a big enough raft to resist the onslaught of seas. If the difference were one of price, we could only advise the pleasure boater, whose chief concern is not just to conform to administrative rules but rather to have a really effective survival device, to buy a Class I life raft.

Yet even in that case, it would be necessary to add to the equipment if one were to be prepared for any circumstance, which would include a survival kit containing, still in Class I, water, food, two parachute flares, and a water pump.

The food rations should be as compact and energy-giving as possible. There should be one or, still better, two ten-liter jerry cans that one should take care not to fill more than halfway so they will float if they fall into the sea. Let us not forget that Catherine's and Lucien's lives were undoubtedly saved because one of the jerry cans was partly empty. Had it been full, it would have sunk like the other one during the first capsize. And why not provide bait? For example, there are tinned mussels that, used as bait, would provide in fish caught a dozen times their own food value.

If one has a Class II or IV life raft, the survival kit becomes naturally more important. In fact, it should contain everything lacking in Class I and also everything lacking in the other classes in relation to Class I. Above all, there should be a signaling mirror, which is the only really effective way of attracting attention during the day.

In the event of an emergency, once the raft is inflated, one need only put aboard the survival kit, as much warm clothing and as many blankets as possible to minimize the hardships of survival. And since it is most unlikely that you will ever have to abandon your boat, you will carry your survival kit with you for years, which you will be sure to inspect or to have inspected periodically. Safety and peace of mind are well worth this small effort of organization.

But take care: the steps one must take before boarding an inflated raft are less simple than one may suppose. First of all, the raft, in its protective cover or container, must be properly stowed on deck, secured to the boat by a stout piece of line about twenty meters long, and it must be easily launched. In Class I and II, inflation is triggered by the line attaching the raft to the boat and only after the raft has been thrown into the sea.

In Class IV, the CO_2 cartridge has to be opened by turning the valve *before* throwing the raft into the sea. No easy task, especially if the valve has been tightly closed. As soon as one hears the gas hissing, toss the whole thing overboard.

Inflation takes place very quickly in every case—in less than a minute. But there's a good chance of the raft landing upside down in the water. The only solution in that case is to dive in and right it. Theoretically, one man using the handles on the raft bottom can do it alone. But perhaps it is wiser for two people to do it. It would be surer, quicker, and less tiring. We have seen how hard it was for Lucien and Catherine.

Never make this mistake: never inflate the raft before launching it. That is what Lucien did and it was, in our opinion, perhaps his only serious mistake. Who knows what damage the raft sustained during the few moments it had to be protected from harm and dragged, without the possibility of taking great care, from the forward deck to the cockpit? Didn't the tiny tear in the raft's bottom that later developed into a permanent leak happen during that feverish moment?

Distress Signals

The most powerful are the *red parachute flares* required by French law aboard all vessels navigating in the

1st, 2nd, and 3rd categories and aboard inflatable life rafts. These flares must reach a height of at least 180 meters. Their visibility is in the range of fifteen miles. It was this type of flare that Lucien launched that Friday night to attract the attention of the freighter that passed not far from the raft.

The freighter was not identified and we pondered at length the reasons which must have prompted the captain not to resume the search at dawn. Did he question the existence of the signal in the first place, a signal never repeated? Did he think that, after having stood by several hours, the absence of a further signal indicated that the boat in distress (he had no way of knowing he was dealing with shipwreck survivors) had managed to resolve its own difficulties?

In any case, the episode is most instructive and justifies, among other things, the necessity of having four parachute flares aboard rafts (one must have four aboard a boat, so why only two on a raft?). The purpose of the parachute flare is to give the alert. Once it has been launched, the rescuers' efforts must be guided by another distress signal—hand-held automatic red fuses (the sticks of phosphorous mentioned by Lucien are not approved in France). Agreed, but that is not the way it always happens. We have just seen proof of that.

To begin with, the first flare may not be noticed. Generally, that happens because the flare has been quite wasted by being launched too far from the coast and without first determining whether there was a reasonable chance of anyone seeing it. Yet, sometimes, as we have seen, the flare goes unnoticed even by those to whom it is directed. In other cases, also, it may be necessary to send off a second flare before being able to make worthwhile use of the red hand fusees.

And giving the alert does not necessarily mean being rescued. We have seen, as in the case of *Njord,* rescuers looking for survivors for hours, not finding them, and these survivors being saved subsequently by other rescuers. The main piece of advice to give about using distress signals is not to be hasty. One must launch them only on the basis of reasonable expectation, when, in one's opinion, the most favorable conditions are united to provide for them the maximum effectiveness. Using them during the day is usually a needless waste.

Incidentally, using them is not always without danger. Catherine has spoken to me of one of her friends—whom I have since met—whose hand was blown off by a parachute flare. She could not help remembering that accident when Lucien burned himself while igniting the flare.

There are many cases of burns caused by distress signals. In my opinion, there are two factors responsible for the occurrence of this type of accident. First, although many flares are manufactured, they are (happily) seldom used. Most of them end their short-lived careers at the bottom of the sea (where they are required to be thrown once their expiration dates have been reached). We are talking, therefore, about things which are seldom if ever handled. If it were otherwise, the manufacturers would soon redesign them to appreciably increase their safe handling.

Next, the person using one is not only sure to be under stress but also nearly always a novice in the matter. Most often the person has just read the instructions for their use for the first time or, if he has read them before, he has surely forgotten them. As things stand now, such a predicament is virtually inevitable. For obvious reasons, it is strictly forbidden to set off distress signals without cause, which at once rules out any experimentation. Would it be possible to do things differently? To authorize experimental launchings in certain areas and under certain conditions? The question has been asked several times. Until now, there has been no official reply.

Red hand fuses are nothing more than powerful Roman candles one must hold at arm's length while they burn. They are visible for five miles. The risk of burning one's hand is less great than with the flares, but the heat of the casing may become so great as to force one to let go before the fuse has fully burned. If one had a glove, it would come in handy (*if* the signal were given aboard a boat, for it would be most unlikely to have such an accessory aboard a raft).

But even if the hand is relatively protected, the same does not hold true for the surroundings. Blown back by the wind, sparks may burn faces, damage sails or, worse still, the rubber raft in which one has taken refuge. Therefore, one must position oneself so that the wind carries the smoke seaward.

There is no danger of burns with smoke signals. Once the pin has been pulled and they have been thrown into the sea they give off a thick red smoke for at least four minutes. Yet there is another risk: that they do no good at all unless used during a dead calm. The slightest breeze blows the smoke along over the sea and dissipates it. Apparently, floating smoke grenades are useful for attracting the attention of an airplane. I have never had any personal experience with them and have never met anyone who could give me any actual information on that point.

For boats navigating in the fourth category (less than twenty miles from shelter), parachute flares are replaced by star flares that release, at an altitude of eighty meters, two red stars whose intensity must be at least 5,000 candlepower for four seconds (as opposed to 10,000 candlepower for thirty seconds at 180 meters for parachute flares). Star flares play exactly the same role of giving the alert as do parachute flares. Less bulky and less heavy, they are visible for approximately ten to twelve miles. Their use aboard boats making short journeys seems to be justified.

What is lacking in this arsenal is a green or white flare enabling one to request help without signaling actual distress. For example, requesting a tow in the event of engine failure during fair weather. Of course, the adoption of a new type of signal entails international conventions, yet that does not mean it cannot be done.

After Abandoning Ship

Once the survivors have abandoned ship and are aboard the inflated raft, they are confronted by this question first. Is it better to try to remain where they are or to drift in the hope of reaching some shore? In neither case, of course, is the possibility of being picked up by a passing ship overlooked.

It all depends on the circumstances. If on the open sea and you have been able to send a radio distress signal giving your position, or if you think parents or friends, worried over your not showing up, will give the alarm and *can furnish sufficiently accurate information to guide the search,* there is no need to hesitate. You must stream the sea anchor and patiently await your rescue, well aware that it may be some time before you are picked up. Too much optimism on that score may lead to an unwisely rapid consumption of provisions and to disappointments that may give way to despair with sometimes fatal consequences. Catherine's great strength lay in having sensed from the beginning that her ordeal would last a long time and that to survive she would have to draw on the limits of human resistance.

But if you feel that no one will worry about your absence for a long time or that no one will know, within fifty miles or so, exactly where to look for you, then there is not much point in staying where you are—unless, of course, you are in a busy shipping lane and you would rather (with some justification) not leave it.

There may be another reason for streaming the sea anchor, at least for a while: heavy weather. Why hadn't Catherine and Lucien tried to steady their raft during the storm by streaming the sea anchor? That was a point I was never really able to clear up. It seems that initially, during the night, after having abandoned *Njord*, they were not aware of having one on board. Moreover, at that point they felt safe and, thanks to physical fatigue and psychological dejection, had but one idea—to sleep. The following morning when they were able to take stock of the raft's equipment, they overlooked the possibility of using the sea anchor for unclear reasons, perhaps because they wanted to cover ground. They actually used the sea anchor only once—to try to remain motionless during the night spent off Majorca. Yet it is legitimate to wonder whether the raft would not have behaved better during the storm behind a sea anchor. It is true that the lack of a sturdy mooring point raises the question of how long the anchor would have remained attached to the raft in such heavy weather and of what damage the strain might have done to the inflatable rings. Yet Lucien and Catherine never mentioned that last reason to me. One thing is certain. Had they known about the sea anchor Friday night, they could have made good use of it to stay near the ship that was searching for them.

Catherine clung tenaciously to the hope of reaching land. She thought it was their only real chance of being saved. In this she was mistaken, for, after having failed to land, Lucien and she were finally rescued by a boat. They were very fortunate. I am convinced that a landing on Majorca in their weakened condition might have had fatal consequences.

For those in peril on the sea, land, quite naturally, exerts such a powerful fascination that it may cause them to make serious mistakes. Take for example the crew of a capsized dinghy who try to swim ashore in vain. One has to accept the fact that the best hope of rescue for shipwreck victims is a boat that will take them to a real port where they can at once receive the care they need.

Catherine was rightly fearful of mishaps that may occur when being picked up, often under perilous conditions, by a ship. And there is no doubt that if the raft had been found by the freighter the first night, a successful rescue could not have been taken for granted. Yet at the same time she grossly underestimated the difficulties and dangers of a landing: a forbidding coast where the risk of drowning was great; cliffs too steep for men and women at the end of their strength to scale or go round; the need, if all went well, to cover a great distance on foot before finding help; coastal currents running out to sea . . .

No, land is not necessarily hospitable to shipwreck survivors. Once again, one must try above all to attract the attention of a ship, and to do so one must, if the physical condition of the raft's occupants permits, set up a continuous watch to avoid missing such an opportunity.

True, Catherine and Lucien did not do so. But Catherine and Lucien had no distress signals left, which practically reduced to nothing their chances of being seen at night. In that case, it was just as well to sleep.

While awaiting rescue—by land or sea—and to preserve the possibility of such an event, to be able to await it as long as possible, there remains the struggle against four foes: cold, fear, thirst, and hunger, with the prospect of exhaustion and despair if one fails to win this struggle.* Two of these foes are particularly awesome: cold and thirst . . . although panic may pose the greatest danger. But that is a psychological manifestation of a person's character which is almost impossible to curb. One can recommend keeping calm and cool, yet such advice is not worth much.

First, there is the cold. Keeping warm is the most pressing concern (if one assumes that the proper functioning of the raft and a successful evacuation of the boat have eliminated, barring further accidents, the danger of drowning). Catherine and Lucien were confronted by truly warm water (20 to 21°C.). Above 20°C., the chances of survival, even in cases of prolonged immersion, are considerable. Five degrees less could have spelled disaster. U.S. Navy research shows that death can occur within minutes in 0°C. water and occurs in every case before one hour. At 5°C., one can survive between half an hour and three hours; at 10°C., between one and six hours; at 15°C., between two and twenty-four hours; at 20°C., between three and forty hours. At 25°C. and above, cold is no longer a problem.

One of the most effective protections against the cold (aside from clothing and blankets) is to have a raft with

---

* As has been demonstrated particularly by Dr. H. Tanguy, president of the French Sailing Boat Yachting Federation's medical commission, in a study published in the magazine *Bateaux* (No. 143, April 1970).

properties of insulation: a double bottom enclosing a gas or cellular material and a tent composed of double or triple layers. We refer the reader here to our reflections concerning the difference in construction requirements between Class I rafts and the others.

Thirst, along with drowning, was certainly the danger that threatened Catherine and Lucien most directly. Without the two downpours that allowed them to replenish their drinking supply, they would surely have perished. They could have tried drinking seawater as Alain Bombard suggests (two swallows eight to ten times a day). Yet the fact is that they never thought of doing it. On the other hand, the rainwater they collected was somewhat briny, having trickled down the salt-crusted tent and the sides of the raft. According to Bombard, the presence of salt in that water was beneficial to the shipwreck victims.

Of course, it is possible to find in the meat of fish enough fresh water to survive in seas that abound with fish, but that is certainly not the case in the Mediterranean. Survivors of shipwrecks in that sea have a very slim chance of finding both food and drink in their catch. Moreover, Lucien and Catherine had nothing, even improvised, that they could use to catch anything. Yet even though they suffered greatly from hunger, their experience shows again that hunger is not the most terrible foe. What they demonstrated above all was that in a life-and-death struggle, stripped to the very bones under the most dramatic circumstances, their most potent weapon was the passion to survive which they so amply displayed. Can we reproach them for having let this fervent desire to escape death lead them—in a state of semidelirium—to the brink of cannibalism? Several other analogous experiences have definitely shown that at a certain level of misery man cannot avoid that pattern. I admire without cavil the freedom of thought they showed by agreeing to bear truthful witness on this point as on all the others.

When I knew them, in November of 1972, they still bore the marks, physical and emotional, of their ordeal. Six months later, they had completely recovered their strength and were emotionally quite stable again. They both told me that neither had ever relived in their dreams the slightest episode of their twelve days of survival. On the other hand, they often dream of *Njord* as she was before the storm as they prepared for their departure. What a marvelous capacity for remembering only happy days! As one might have foreseen, they are again on good terms with the sea that, nevertheless, dealt with them so cruelly. Sailing is an ever more intimate part of their existence. Perhaps you will meet them some day in a chance encounter as you call on some port. Their names are Catherine and Lucien.